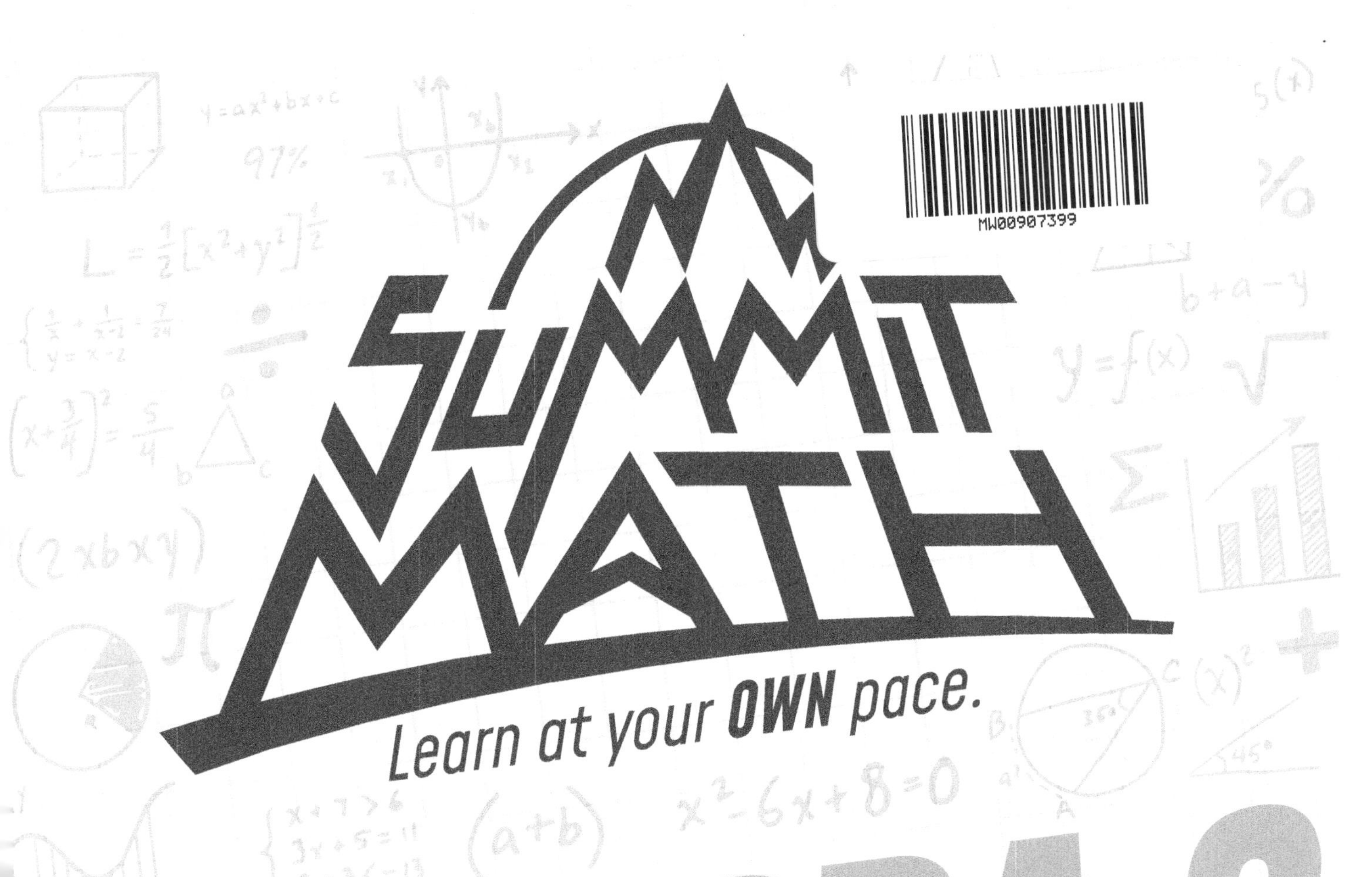

ALGEBRA 2

second edition

1 INTRODUCTION TO FUNCTIONS

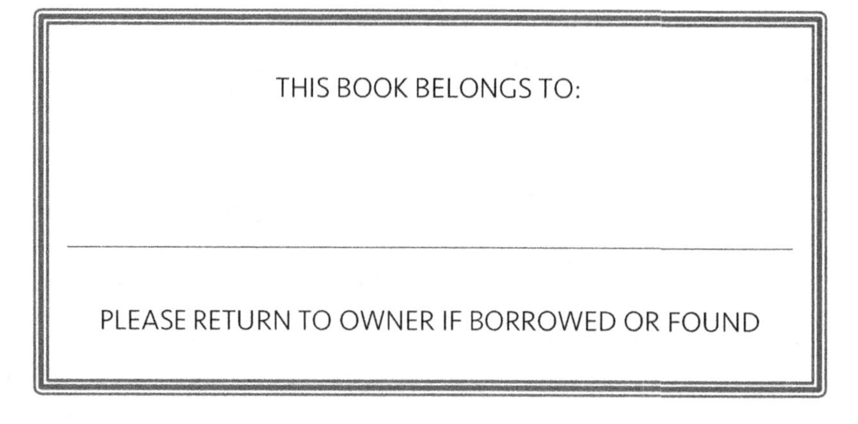

DEDICATION

To Lauren, Chloe, Dawson and Teagan

ACKNOWLEDGEMENTS

I started writing these books in 2013 to help my students learn better. I kept writing them because I received encouraging feedback from students, parents and teachers. Thank you to all who have used these books, pointed out my mistakes, and made suggestions along the way. Thank you to all of the students and parents who asked me to keep writing more books. Thank you to my family for supporting me through every step of this journey.

This book was typeset in the following fonts:
Seravek + Mohave + *Heading Pro*

Graphics in Summit Math books are made using the following resources:
Microsoft Excel | Microsoft Word | Desmos | Geogebra | Adobe Illustrator

First printed in 2017

Printed in the U.S.A.

Summit Math Books are written by Alex Joujan.

www.summitmathbooks.com

Learning math through Guided Discovery:

A Guided Discovery learning experience is designed to help you experience a feeling of discovery as you learn each new topic.

Why this curriculum series is named Summit Math:

Learning through Guided Discovery can be compared to climbing a mountain. Climbing and learning both require effort and persistence. In both activities, people naturally move at different paces, but they can reach the summit if they keep moving forward. Whether you race rapidly through these books or step slowly through each scenario, this curriculum is designed to keep advancing your learning until you reach the end of the book.

Guided Discovery Scenarios:

The Guided Discovery Scenarios in this book are written and arranged to show you that new math concepts are related to previous concepts you have already learned. Try to fully understand each scenario before moving on to the next one. To do this, try the scenario on your own first, check your answer when you finish, and then fix any mistakes, if needed. Making mistakes and struggling are essential parts of the learning process.

Homework and Extra Practice Scenarios:

After you complete the scenarios in each Guided Discovery section, you may think you know those topics well, but over time, you will forget what you have learned. Extra practice will help you develop better retention of each topic. Use the Homework and Extra Practice Scenarios to improve your understanding and to increase your ability to retain what you have learned.

The Answer Key:

The Answer Key is included to promote learning. When you finish a scenario, you can get immediate feedback. When the Answer Key is not enough to help you fully understand a scenario, you should try to get additional guidance from another student or a teacher.

Star symbols:

Scenarios marked with a star symbol ★ can be used to provide you with additional challenges. Star scenarios are like detours on a hiking trail. They take more time, but you may enjoy the experience. If you skip scenarios marked with a star, you will still learn the core concepts of the book.

To learn more about Summit Math and to see more resources:

Visit www.summitmathbooks.com.

As you complete scenarios in this part of the book, follow the steps below.

Step 1: Try the scenario.
Read through the scenario on your own or with other classmates. Examine the information carefully. Try to use what you already know to complete the scenario. Be willing to struggle.

Step 2: Check the Answer Key.
When you look at the Answer Key, it will help you see if you fully understand the math concepts involved in that scenario. It may teach you something new. It may show you that you need guidance from someone else.

Step 3: Fix your mistakes, if needed.
If there is something in the scenario that you do not fully understand, do something to help you understand it better. Go back through your work and try to find and fix your errors. Mistakes provide an opportunity to learn. If you need extra guidance, get help from another student or a teacher.

After Step 3, go to the next scenario and repeat this 3-step cycle.

NEED EXTRA HELP?
watch videos online

Teaching videos for every scenario in the Guided Discovery section of this book are available at www.summitmathbooks.com/algebra-2-videos.

CONTENTS

Section 1

COMPARING INDEPENDENT AND DEPENDENT QUANTITIES

The word "function" can be used in a wide range of contexts. A broken machine does not function properly. It malfunctions. The function of a light switch is to turn a light on or off. Every time you use a computer or calculator, you make it use a function to perform a task.

1. In a mathematical context, a function is something that shows how one quantity affects another quantity. When a function relates two quantities, it is common to call one of them dependent, because its value depends on the value of the other quantity. Identify the dependent quantity in each statement below.

    a. A child's height is a function of the child's age. (changes with)

    b. The area of a circle is a function of the length of its radius. (is determined by)

    c. The total amount of calories burned during exercise is a function of the time that a person spends walking. (depends on)

2. If a person's paycheck, P, is a function of the number of hours worked, h, you can quickly refer to the paycheck function as P of h, or $P(h)$. This notation will be explained more in the next section. In each statement below, name the function using the two variables.

    a. A child's height, H, is a function of the child's age, a. The height function is _____.

    b. The area of a circle, A, is a function of the length of its radius, r. The area function is _____.

    c. The total amount of calories burned during exercise, C, is a function of the time that a person spends walking, t. The calorie function is _____.

3. Read the description of each experiment. After you read the experiment, identify the dependent variable and the independent variable in the experiment.

    a. Fill several buckets with water, and heat each bucket to a different temperature. Measure the amount of sugar that dissolves in each bucket of water.

    b. Light a candle and measure its height every minute.

    c. Make a bridge with a piece of paper and measure the number of pennies you can stack up before the paper collapses. Add another layer of paper and stack pennies again. Observe how the number of layers of paper relates to the number of pennies you can stack.

Use this page to record important ideas in the previous section or for any other writing that helps you learn the topics in this book.

Section 2

REPRESENTING A FUNCTION WITH AN EQUATION OR A GRAPH

There are many ways to represent a function, or to show how one quantity affects another quantity. Two common ways to display a function are through an equation or a graph.

4. Consider the function represented by the equation $F=10x+3$. In this equation, F is a function of x because the value of F depends on the value of x.

   a. If $x=0$, then $F=$ _____ .

   b. If $x=4$, then $F=$ _____ .

   c. For the function $F=10x+3$, which variable is the dependent quantity?

5. Consider the function represented by the graph shown. The graph shows how the cost to rent a cabin for a night depends on how many people will sleep in the cabin.

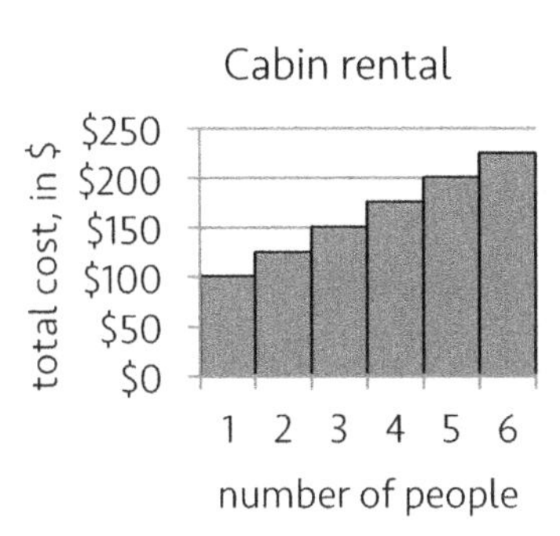

   a. How much does it cost for 3 people to rent the cabin?

   b. For the function shown, identify the independent quantity.

   ★c. Write an equation that shows the total cost, C, if *n* people rent the cabin.

6. Bamboo plants are known for growing quickly. Suppose you buy a bamboo plant, bring it home, and measure its height each day. As it grows, the height of the bamboo is a function of the number of days you have owned it. This function can be shown by an equation. Suppose the height function is $H=25+2d$, where H is the height, in inches, after you have owned the plant for d days.

   a. How tall is the bamboo after you have owned it for 10 days?

   b. How much does the bamboo's height increase each day?

   c. What does the 25 represent in the height function, $H=25+2d$?

7. The height function can also be shown in a graph, with the two quantities shown on the two axes.

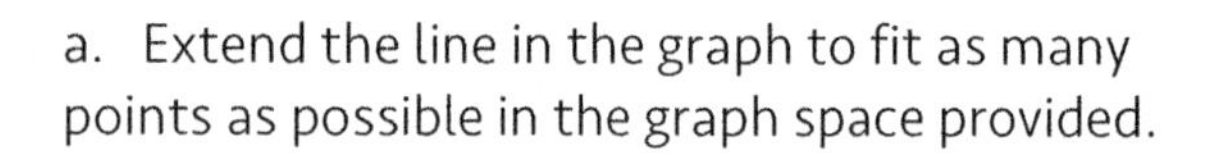

a. Extend the line in the graph to fit as many points as possible in the graph space provided.

b. Using the graph, after how many days will the bamboo reach a height of 105 inches?

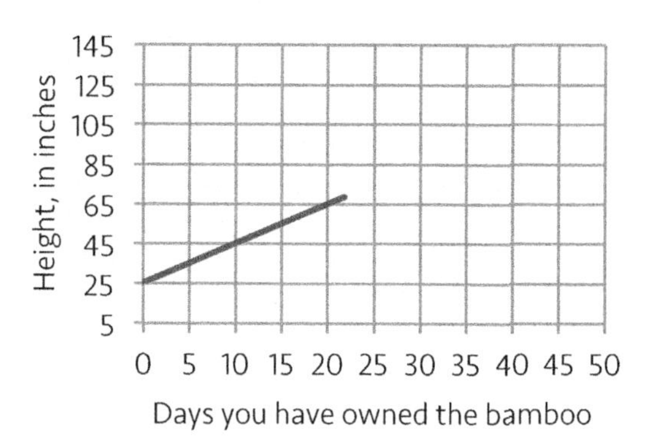

8. ★Some information from the height function is shown in a table, with the columns labeled to identify the two quantities.

a. Fill in the missing value in the table.

b. Using the table, how many days does it take the bamboo plant to grow 120 inches taller?

days, d	height, H
0	25
20	65
40	105
60	145
80	185
100	

9. In the previous scenario, the height, H, is a function of the number of days, d. Using only the variables, you can say that H is a function of d. Use more concise terms, you can refer to the height function as "H of d." For example, the height of the bamboo is 225 inches after 100 days. Using more concise terms, you can state that H of 100 is 225.

a. What is H of 80? ("What is the height of the bamboo after 80 days?")

b. What is H of 150?

c. What is the value of d, if the value of H of d is 125?

10. The expression H of d is written as $H(d)$ by mathematicians. With this notation, the question "What is H of 80?" can be written as "What is $H(\_\_\_)$?". Fill in the blank.

11. Rewrite each statement in more concise function notation. The first one is started for you.

a. The height is 31 inches after 3 days. $\rightarrow H(3) =$ _____

b. The height is 47 inches after 11 days. $\rightarrow$ ____________

c. After 27 days, the height is 79 inches. $\rightarrow$ ____________

12. As a reminder, the height function was defined earlier by the equation $H = 25 + 2d$. The notation $H(d)$ shows that the height, *H*, is dependent on *d*. The height is the output, while *d* is the input.

a. Using the equation $H = 25 + 2d$, what is $H(1)$?

b. What is $H(115)$?

c. What is the value of *d* for which $H(d) = 87$ inches?

d. Why is it that $H(0) = 25$?

13. Suppose you buy another bamboo plant and its growth is shown by the function $H = 10 + 1.9d$. Once again, a shorter way to refer to this function is to call it $H(d)$.

a. Using words, what is the meaning of $H(13)$?

b. Does this bamboo plant grow faster or slower than the first one you bought? Explain.

14. The graph shows data for a cyclist during a 120-mile race. The variable *M* represents "Miles traveled", while *H* is "Hours of riding." Since the miles traveled is a function of the hours of riding, the function can be referred to as *M* of *H*, or $M(H)$.

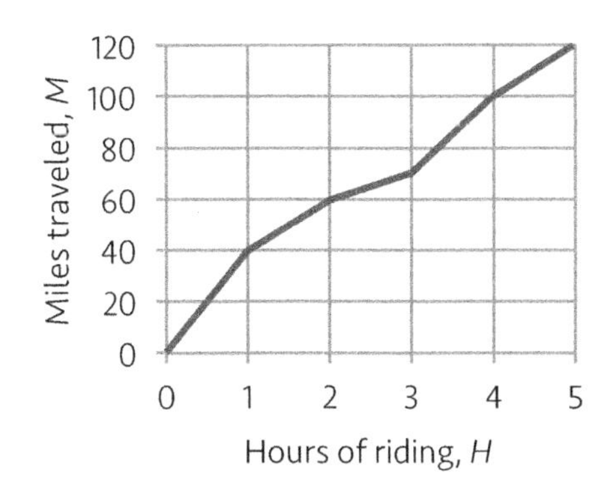

a. How far has the cyclist traveled after 2 hours?

b. How long does it take the cyclist to travel 100 miles?

c. Estimate the value of $M(3)$.

d. For what value of *H* does $M(H) = 120$?

15. Use the previous graph to analyze the speed of the rider during various portions of the race.

a. Estimate how fast the cyclist was riding during the first hour of the race.

★b. Estimate the cyclist's speed during the second hour of the race.

★c. Estimate the speed of the cyclist during the 3rd hour of the race (the time between the 2nd hour and the 3rd hour), in miles per hour.

16. The graph displays data from another long bike ride, where the distance, $D(t)$, is a function of the time t. The entire ride is shown in the graph.

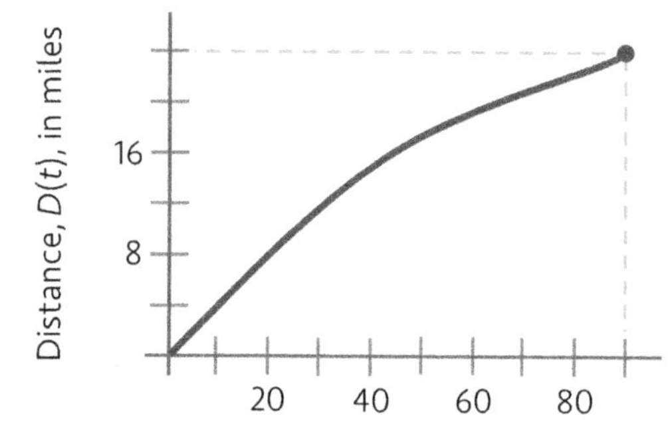

a. Estimate the value of $D(20)$.

b. Estimate the value of t for which $D(t)=16$.

★c. During which portion of the ride did the rider maintain the fastest average speed? What was the average speed during this portion?

17. The graph of the function $R(n)$ is shown.

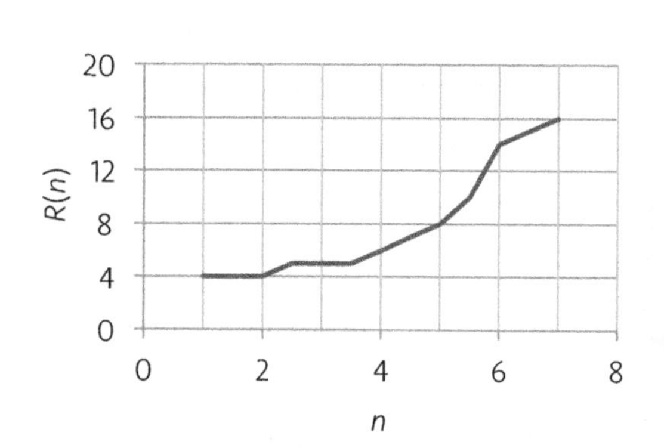

a. What is the value of $R(5)$?

b. What is the value of n if $R(n) = 16$?

As a reminder, a function shows the relationship between two quantities. If you think about these quantities as x's and y's, then the first question in the previous scenario involves finding a y-value when you are given a specific x-value. In the second question, you are given a y-value and you are asked to find the x-value that is paired with it.

18. Consider a function given by the equation $y=2x+5$.

a. What is the value of y if $x = 3$?

b. What is the value of x if $y = -11$?

19. Consider the function given by the equation $V(e)=9e+2$.

a. You may be more familiar with having the variables x and y in your equations. For the function $V(e)=9e+2$, which variable is taking on the role of "x" in this equation?

b. Which variable is "y" in the function $V(e)=9e+2$?

c. The notation $V(e)$ shows that V gets its value from e. What is the value of $V(10)$?

d. What is the value of e, if the value of $V(e)=-7$?

20. A function is defined by the equation $K(g)=7-3g$. Fill in each blank below.

a. $K(-10)=$__________

b. $K($______$)=49$

21. Use the graph to find the value of each expression listed below. Estimate if necessary.

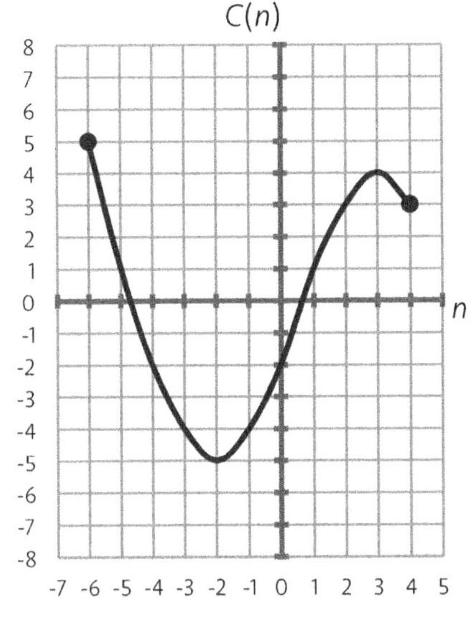

a. $C(3)$

b. $C(-5)$

c. $C(-10)$

d. $C(7)$

22. Use the graph in the previous scenario to find each value below. Estimate if necessary.

a. What is n if $C(n)=5$?

b. What is n if $C(n)=3$?

c. Fill in the blank.

$C(\_\_\_\_)=-4$

d. Fill in the blank.

$C(\_\_\_\_)=-7$

23. The function $f(x)$ is defined by the equation $f(x)=x^2-5$. Fill in each blank below.

a. What is $f(-6)$?

b. What is x if $f(x)=95$?

24. A boy gets up early one morning to go for a long ride on his bicycle. Thirty-six minutes later, his older sister realizes he left the inhaler that he uses for his asthma, so she gets in the car and tries to catch him. Let $B(m)$ represent the boy and $S(m)$ represent the sister.

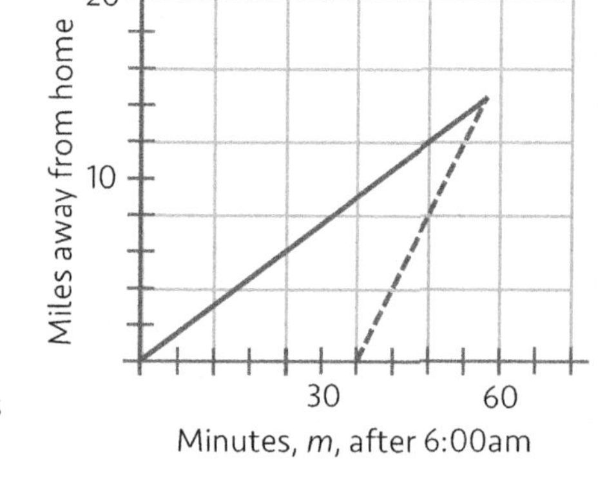

a. Using words, what is the meaning of $B(12)$?

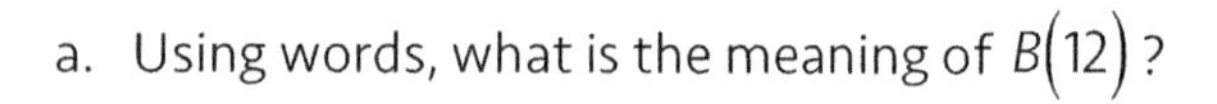

b. What is the value of $B(48)-S(48)$? Using words, what does this value represent in the context of the scenario?

c. Why does $S(36)=0$?

25. ★Use the graph in the previous scenario to answer the following questions.

a. If you find the rate at which the brother moves as he rides his bike, what would you use for the units of his rate if you refer to each axis as it is marked in the graph?

b. How fast does the brother move when he is riding his bike, measured in miles per hour?

c. The sister's rate is ______ miles per hour faster than her brother's rate.

Use this page to record important ideas in the previous section or for any other writing that helps you learn the topics in this book.

Section 3

INCREASING, DECREASING OR CONSTANT

26. Try to fill in the blanks. When a function has a positive slope, the function is described as increasing. When a function has a negative slope, it is described as ________________. When the slope of a function is ___, it is described as constant.

27. A function that only has a positive slope is called strictly increasing, because the y-values always increase as you move along the graph from left to right. An example of a strictly increasing function is a line with a positive slope. Explain why the function shown is NOT strictly increasing.

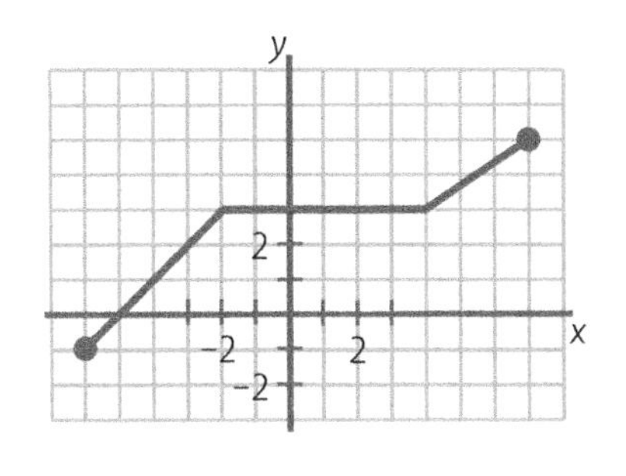

28. The previous function is not increasing in the section between (−2, 3) and (4, 3). In this interval, the function is considered to be constant. The word "constant" is used in this case because the function's y-values are not changing as the x-values increase. The y-values are constant. What can you look for in the graph of a function to find a section where the function is constant?

29. The previous function is constant between 2 points, (−2, 3) and (4, 3). Another way to state this is to focus on the x-values and say that the function is constant between the x-values of −2 and 4. A more concise way to show this is with interval notation: $-2 < x < 4$ (points with x-values between −2 and 4). Write each statement below using interval notation.

a. between the x-values of −5 and 1, but not including −5 or 1

b. between the x-values of 4 and 20, including 4 but not 20

30. Identify the interval(s) on which each function is constant. Write the interval(s) as an inequality.

a.

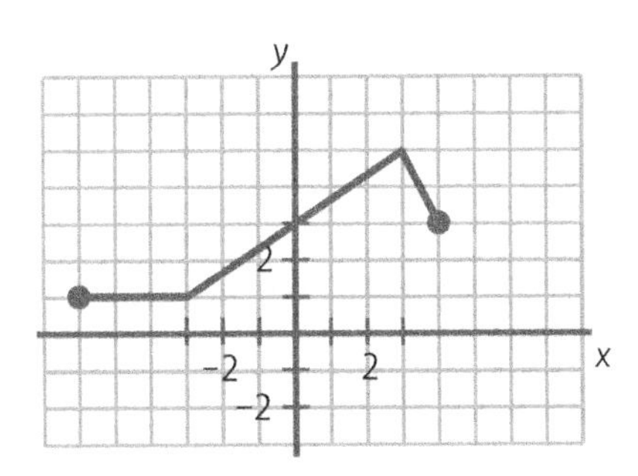

b.

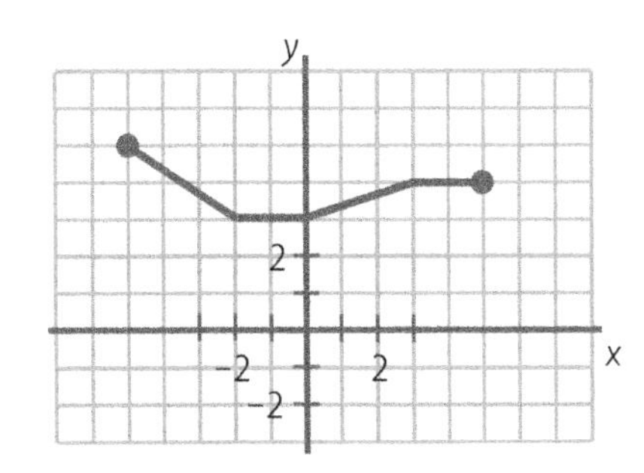

31. Identify the interval(s) on which each function in the previous scenario is decreasing.

32. Describe each function below using the words increasing, decreasing, or constant.

a.

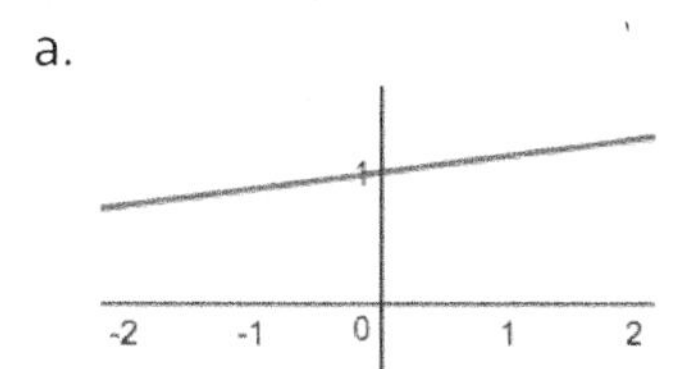

b.

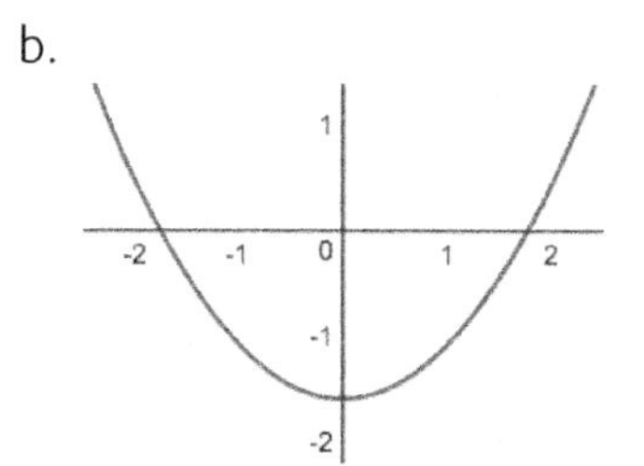

c.

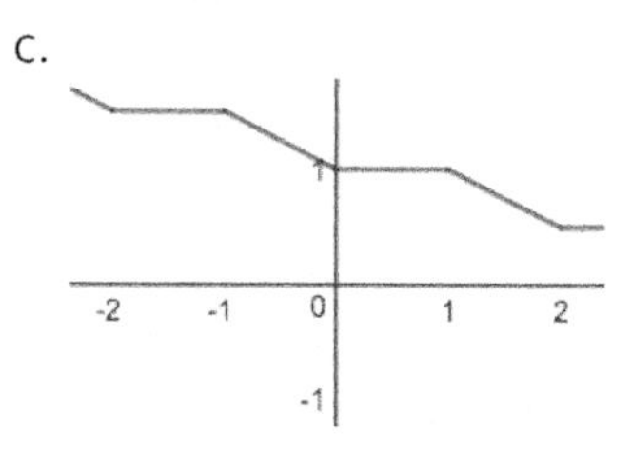

33. The graph of a function $g(x)$ is shown. Determine the value of each expression listed below.

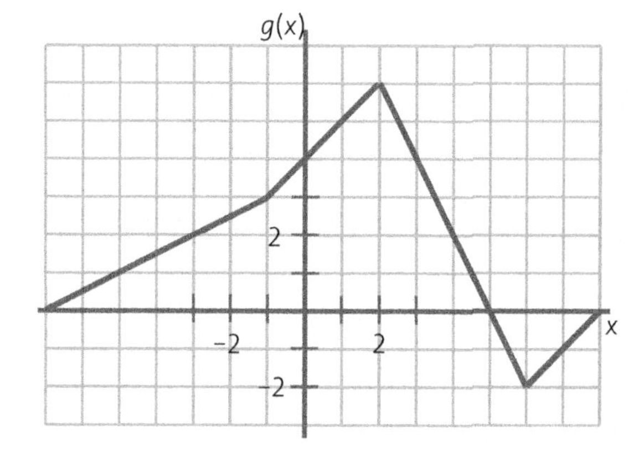

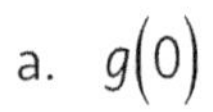

a. $g(0)$

b. $g(-3)$

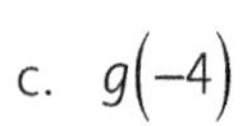

c. $g(-4)$

34. Use the graph of the previous function to find each value below.

a. What is the x-coordinate for the point that has a y-value of 2?

b. What is the value of x when $g(x)=6$?

c. Find x if $g(x)=0$.

d. For which x-value does $g(x)=-3$?

35. For what values of x is $g(x)$ decreasing? Write this as an interval.

36. On what interval(s) is $g(x)$ increasing?

Refer to the graph of the function $g(x)$ to answer the following questions.

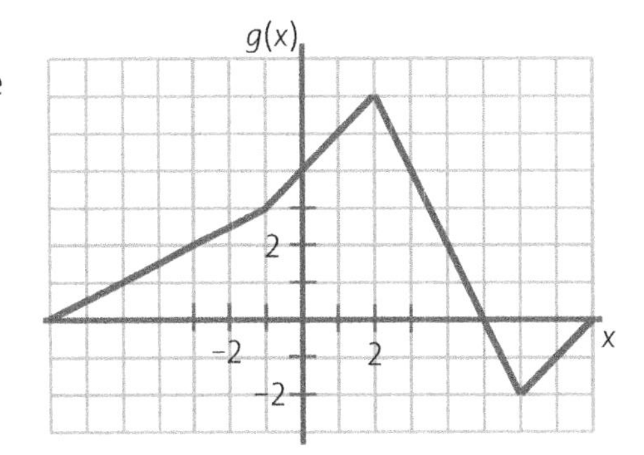

37. Find the ordered pair on $g(x)$ that has the least x-value. What is the x-value of this point?

38. Find the ordered pair on $g(x)$ that has the greatest x-value. What is its x-value?

39. What are the least and greatest g-values shown on the graph of $g(x)$?

Use the graph of $C(n)$ to answer each question below.

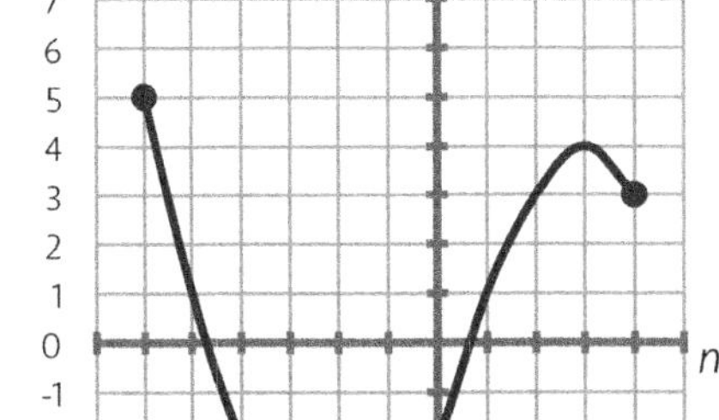

40. On what interval(s) is $C(n)$ increasing?

41. On what interval(s) is $C(n)$ decreasing?

42. What are the least and greatest n-values shown on the graph of $C(n)$?

43. What are the least and greatest C-values shown on the graph of $C(n)$?

44. The function $T(h)$ is defined by the equation $T(h)=h^2-2h-3$.

a. What is $T(-2)$?

b. What is h if $T(h)=0$?

45. Sam gets to work on Monday morning by walking to the bus stop, riding the bus for 15 minutes, and then walking the rest of the way. To get home, he completes this same route in a reverse order. His morning commute and afternoon commute are shown below.

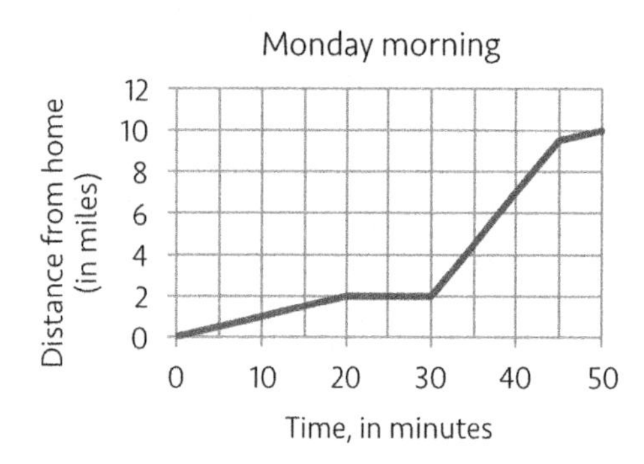

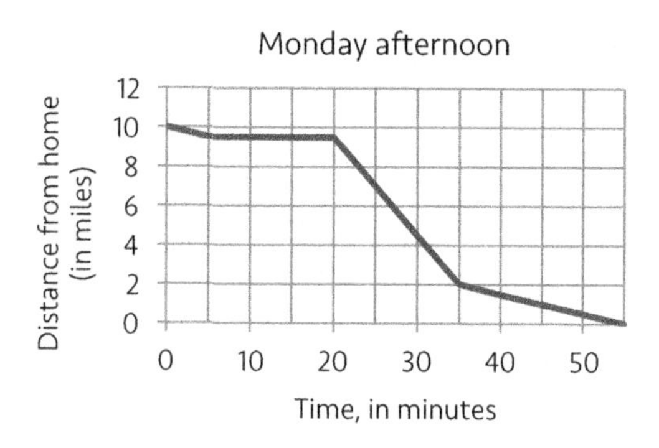

a. Explain why the Monday afternoon commute is not a mirror image of the Monday morning commute.

b. Why is the Afternoon commute longer than the Morning commute?

c. Estimate Sam's walking speed, in miles per hour.

★d. Estimate the speed of the bus, in miles per hour.

46. Consider a function that is called $f(x)$.

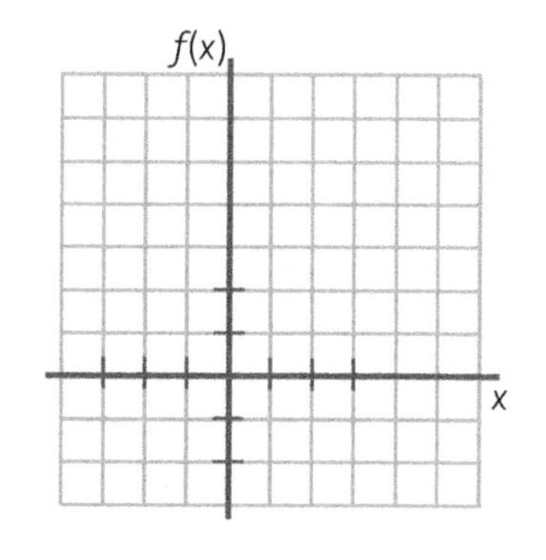

a. The value of $f(1)$ is -2. Show this as a point in the graph to the right.

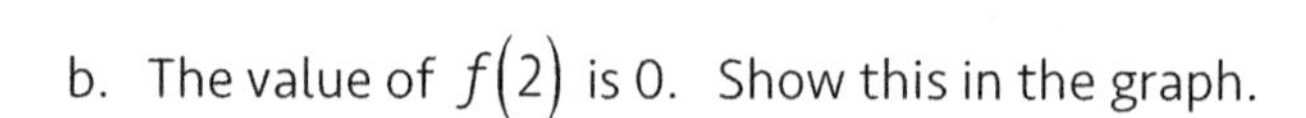

b. The value of $f(2)$ is 0. Show this in the graph.

c. $f(3)=2$. Show this in the graph.

47. In the previous scenario, suppose the pattern of the dots continues.

a. What is $f(4)$?

★b. What is the value of $f(6)$?

48. In the previous scenario, $f(x)$ is a linear function. It is called linear because you can draw a line through all of the points to show that there are more points than the ones you drew. Write the equation for $f(x)$, in Slope-Intercept Form.

Use this page to record important ideas in the previous section or for any other writing that helps you learn the topics in this book.

Section 4

THE DOMAIN AND RANGE OF A FUNCTION

49. Consider the function shown. If you try to list all the points that are part of the function, you will run out of time because there are infinitely many points. You can, however, express that the points are between certain values.

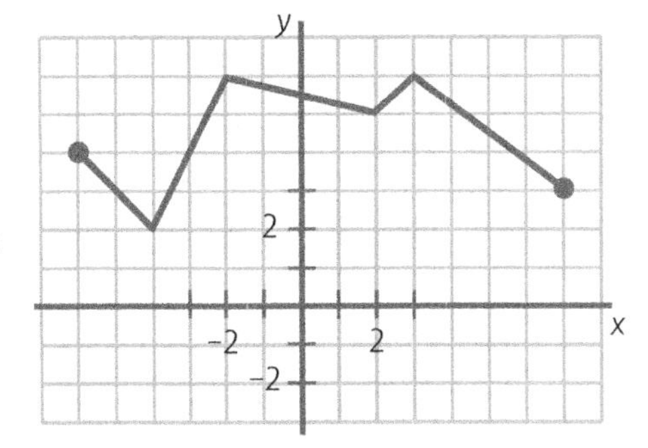

a. What are the least and greatest x-values that are part of this function?

b. What are the least and greatest y-values of the function?

At this point, you are ready to learn two new mathematical words: domain and range. The **domain** of a function is all of the x-values of the function. The domain of the function in the previous scenario is the x-values between −6 and 7. As an inequality, the domain can be written as $-6 \leq x \leq 7$. The **range** of a function is all of the y-values of the function. The range of the function in the previous scenario is the y-values between 2 and 6. As an inequality, the range can be written as $2 \leq y \leq 6$.

50. A function is shown in the graph.

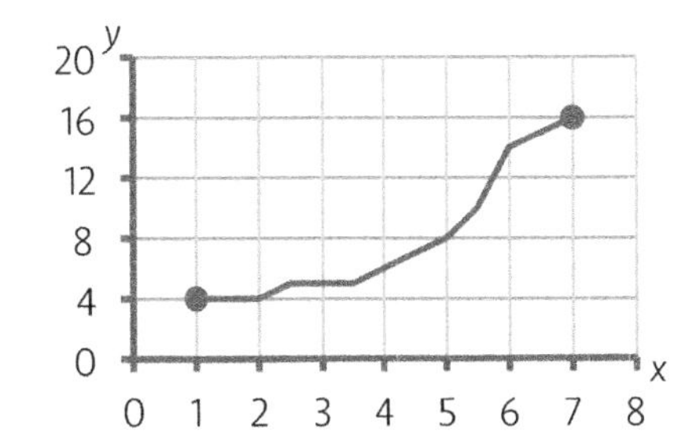

a. What is the domain of the function?

b. What is the range of the function?

51. Identify the domain of each function shown below.

a.

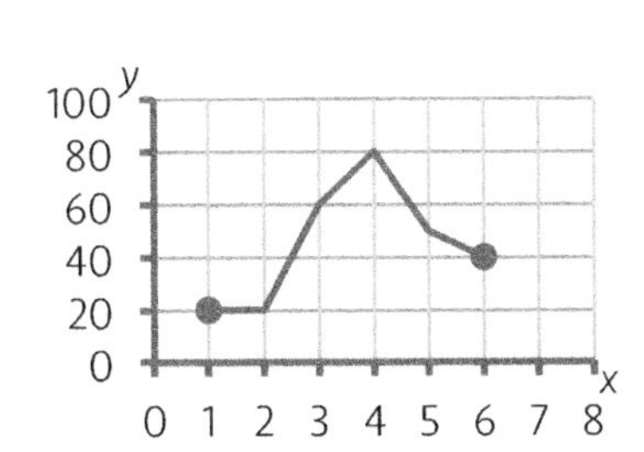

b.

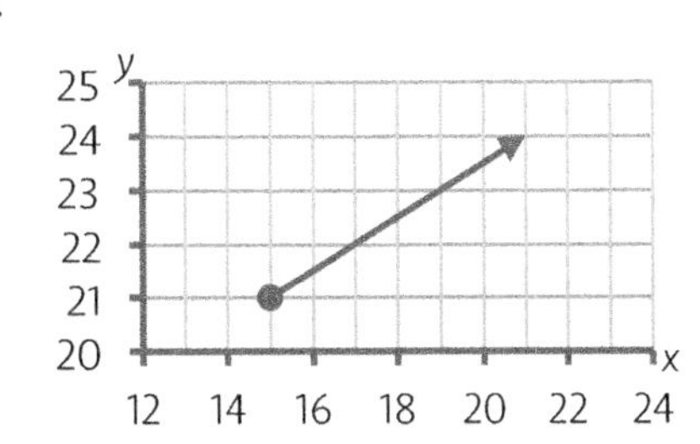

52. Identify the range of each function in the previous scenario.

53. Identify the domain of each function shown below.

a.

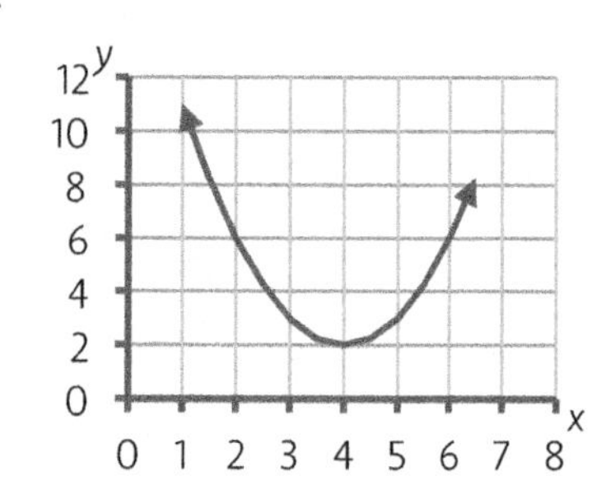

b.

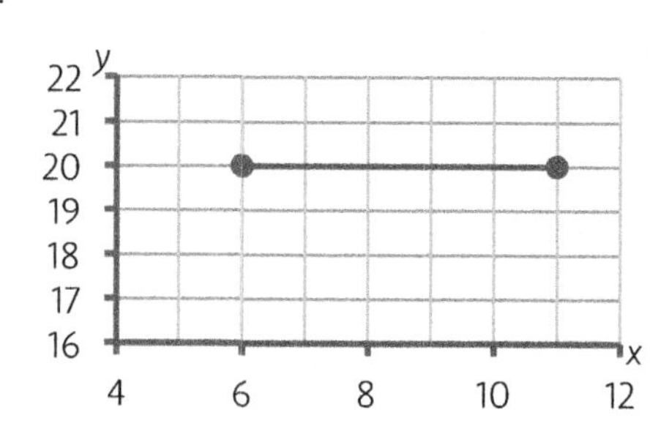

54. Identify the range of each function in the previous scenario.

55. If you graph the line given by the equation $y=4$, it will match the image shown in the graph to the right.

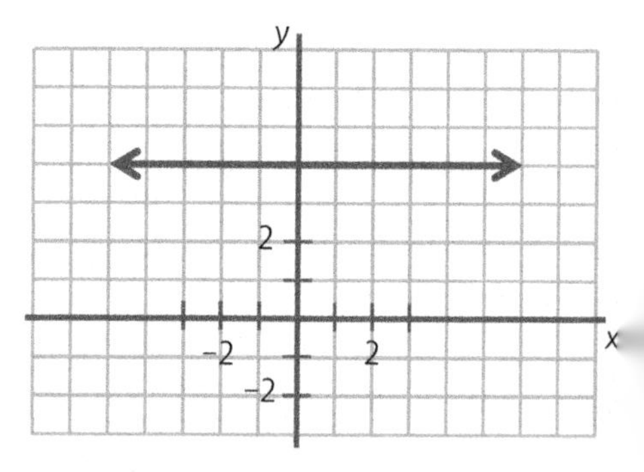

a. Why are arrows drawn at the ends of the line?

b. What is the domain of the line $y=4$ if you graph all of the points that are part of this line, including the points beyond the arrows?

56. If you graph the line given by the equation $y=\frac{3}{4}x+2$, it will match the image shown in the graph to the right. What is the domain of the line $y=\frac{3}{4}x+2$ if you graph all of the points that are part of this line, including the points beyond the arrows?

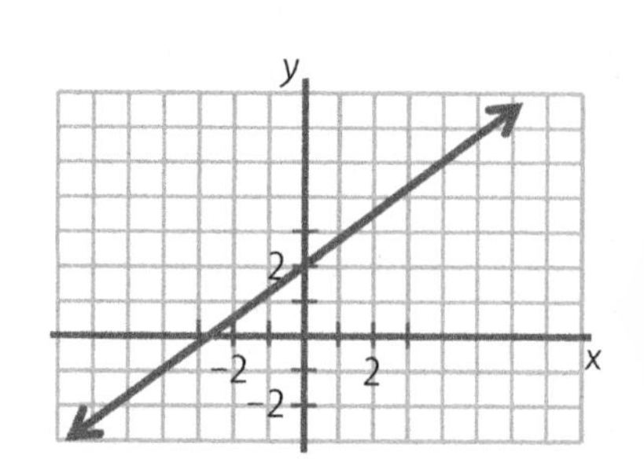

57. The concept of infinity is worth mentioning at this point. How do you define this word? When you read the word "infinity" in a math context, what do think this word means?

When you graph the line $y=\frac{3}{4}x+2$, it extends in both directions without ending. It is not possible to actually draw this, but you can draw arrows to show that the line keeps going. In this sense the domain is uncountable, so you can say that the domain is infinite. Every x-value on the x-axis is part of the function. Another way to show that the domain is every x-value is to state that the domain is "all real numbers."

58. Consider what happens when you make the domain smaller for the line $y=\frac{3}{4}x+2$. Suppose you restrict the domain by only graphing the part of the line that has x-values from −4 to 4. Graph the line if its domain is only the x-values from −4 to 4, including the values of 4 and −4.

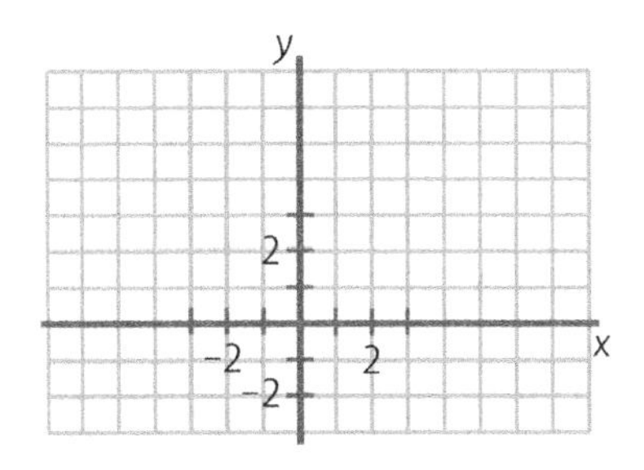

Returning to the concept of infinity, even when you restrict the domain to $-4 \le x \le 4$, there are still an uncountable number of x-values (x could be 1.4, or 1.27, or 1.538, or 1.9462, or...). Once again, the domain is uncountable, or infinite.

59. If you restrict the domain of $y=\frac{3}{4}x+2$ to the x-values between 0 and 1, is the domain still infinite?

60. A Geometry concept is worth mentioning here. By definition, a line extends without ending in both directions. If the arrows are replaced with endpoints and the line is now only the region between the endpoints, the line is now called a ___________________.

61. Consider a scenario that does not have an infinite domain. A coin-sorting machine can be viewed as performing a function. If a coin goes into the machine, its size is measured and it is placed in a numbered slot. For this function, the domain is the coin type (these are the inputs) and the range is the slot number (the outputs). Suppose the coin-sorting function works as shown.

Coin Type	Penny	Nickel	Dime	Quarter
Slot Number	1	2	3	4

a. What is the domain of the coin-sorting function?

b. What is the range of the function?

62. Suppose there is a simple function that only consists of the four ordered pairs shown. State the domain and range of this group of ordered pairs.

$$\{(0,-2), (-5, 3), (0, 6), (-4, 1)\}$$

63. A function is defined by the equation $f(x)=3x+1$. Suppose the domain is restricted to only five values: 0, 1, 2, 3, and 4. Since these values are the restricted domain, they are the only x-values that you can use in the equation. Calculate the value of $f(x)$ for each of the given x-values and combine these values to show the restricted range of this function.

64. Another function is shown below and it is assigned a restricted domain. Use the given domain to find the range of the function.

function: $h(x)=x^2-5x$ domain: $\{-5, -2, 0, 2, 5\}$ range:

65. Consider the given function. It only contains the points shown (including the endpoints).

a. What are the smallest and largest x-values shown in the function?

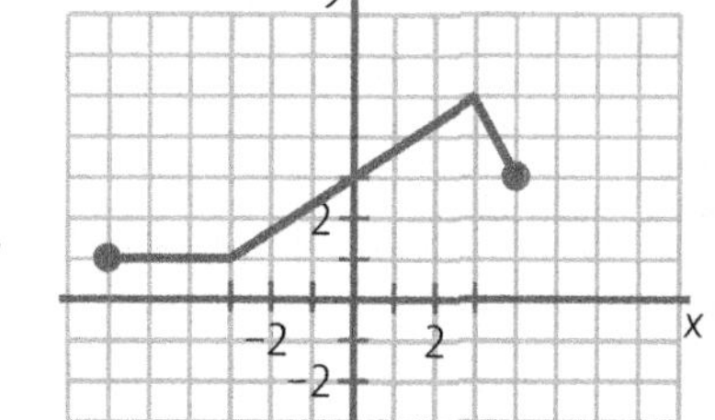

b. The domain of the function is the x-values between ____ and ____. As an inequality, the domain can be written as ____ $\le x \le$ ____ .

c. What are the least and greatest y-values shown in the function?

d. The range of the function is the y-values between ____ and ____. As an inequality, the range can be written as ____ $\le y \le$ ____ .

66. There is not a single equation that will produce the graph in the previous scenario. It has been formed by taking segments from 3 different lines. Write the 3 equations that would represent the 3 lines.

67. If you graphed each of the previous 3 lines that you identified in the previous scenario, the lines would extend in both directions without ending. In the previous scenario, however, only a small portion of each line is shown because they have restricted domains. Identify the restricted domain for each line.

a. The restricted domain for the line given by the equation $y=1$ is ______ $\leq x \leq$ ______ .

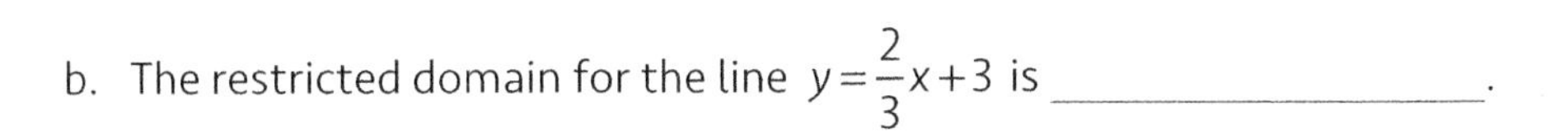

b. The restricted domain for the line $y=\frac{2}{3}x+3$ is ____________________.

c. The restricted domain for the line $y=-2x+11$ is ____________________.

68. Graph the function that is described below. It is made by graphing a small portion of 2 separate lines.

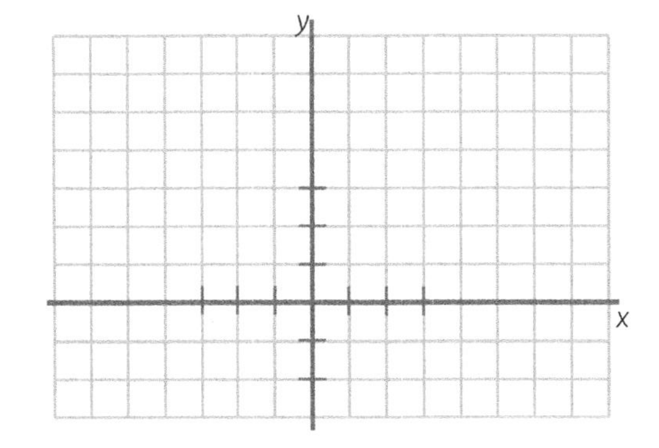

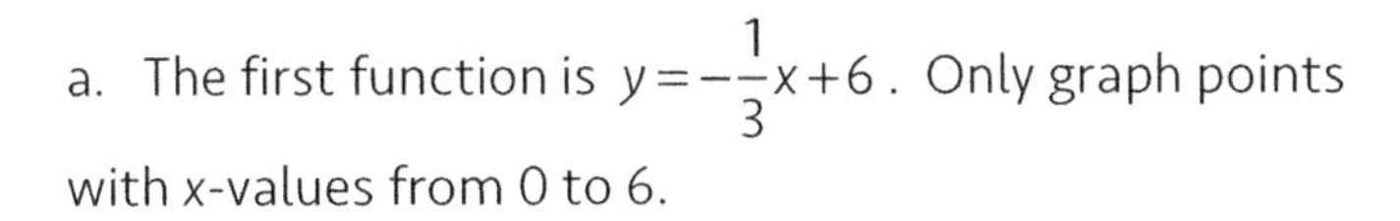

a. The first function is $y=-\frac{1}{3}x+6$. Only graph points with x-values from 0 to 6.

b. The second function is $y=x+6$. Only graph points with x-values from −6 to 0.

69. In the previous scenario, suppose the entire function is given the name $F(x)$.

a. What is the value of $F(3)$?

b. What is the value of $F(-2) + F(0)$?

70. Refer to the graph in the previous scenario to answer the following questions.

a. What are the least and greatest x-values shown in the function?

b. The domain of the function is the x-values between ______ and ______. As an inequality, the domain can be written as ______ $\leq x \leq$ ______ .

c. What are the lowest and highest y-values shown in the function?

d. The range of the function is the y-values between ______ and ______. As an inequality, the range can be written as ______ $\leq y \leq$ ______ .

71. Use the graph of the function $D(t)$ to estimate the following function values.

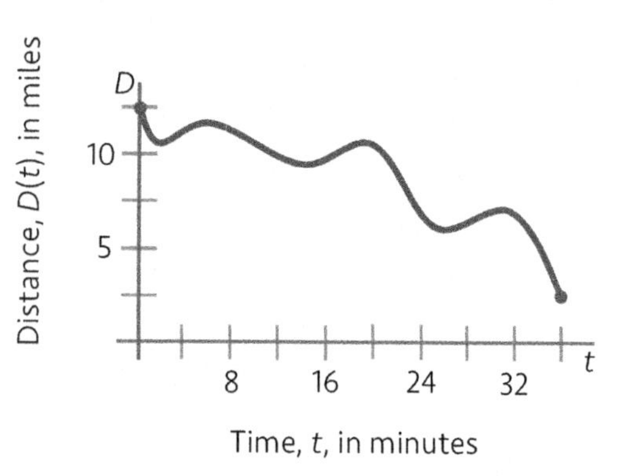

a. 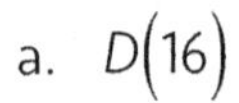 $D(16)$

b. 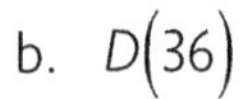 $D(36)$

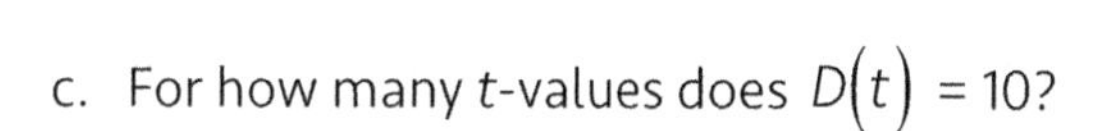
c. For how many t-values does $D(t) = 10$?

72. The previous scenario compares two quantities. Which one is the independent quantity?

73. Using what you have seen so far, when a function is plotted in a graph, which axis is used to represent the independent quantity?

Use this page to record important ideas in the previous section or for any other writing that helps you learn the topics in this book.

Section 5

MORE SCENARIOS THAT INVOLVE FUNCTIONS

74. The temperature is measured over the course of a 21-hour period. In the graph below, when $h = 0$, the time is midnight. When $h = 18$, the time is 6:00pm.

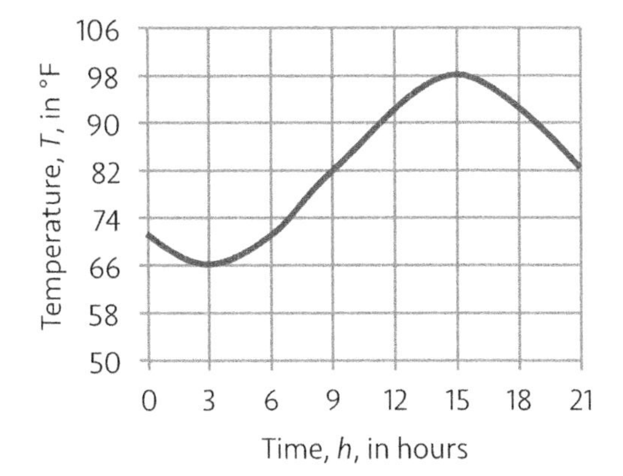

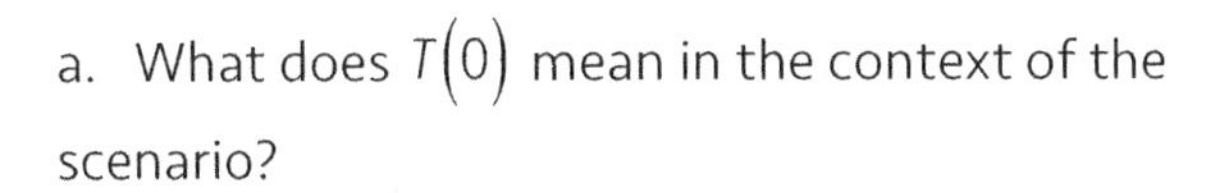

a. What does $T(0)$ mean in the context of the scenario?

b. What is $T(15)-T(3)$ and what does this value represent in the context of the scenario?

c. For what value of h does $T(h)=82$?

d. For what values of h is the temperature decreasing?

75. Refer to the graph in the previous scenario to answer the following questions.

a. What are the least and greatest h-values shown in the function's graph?

b. What are the least and greatest T-values shown in the function's graph?

c. Write the domain of the function as an inequality, if the function is restricted to only what is shown in the graph.

d. Write the range of the function as an inequality, if the function is restricted to only what is shown in the graph.

76. A company pays their employees using the function $P = 10H + 15E$, where P is the total paycheck after H hours of work, until H is more than 40. In the function above, E represents the extra hours worked during the week, if an employee works more than 40 hours. For example, if an employee works 50 hours, he or she is paid an hourly rate for the first 40 hours and then a higher hourly rate for the last 10 hours. Paychecks are given each week. Since P is dependent on two variables, you can refer to this paycheck function as $P(H,E)$.

a. What is the value of $P(40,2)$?

b. How many hours did an employee work during the week if the total paycheck for that employee is represented by $P(40,2)$?

c. What does the 10 represent in the function $P = 10H + 15E$?

d. What does the 15 represent in the function above?

e. How much does an employee earn for 51 hours of work?

77. ★Jan works for the company in the previous scenario. How many hours did Jan work last week if she was paid $700?

78. A taxi in New York City charges an initial fee of $2.50 plus an extra $2.00 per mile. The taxi also considers the impact of traffic and charges $0.40 for every minute that the taxi is stopped in traffic.

a. How much would a taxi charge for a 2.5 mile trip that included 3 minutes of time stopped in traffic?

b. Suppose you use a taxi to travel 4.2 miles and the driver tells you that the total charge for the ride is $13.70. How many minutes was the taxi stopped in traffic?

★c. The taxi's cost function (call it C) depends on two quantities. Let m represent the miles traveled and t represent the time stopped in traffic. Since C is dependent on two quantities, you can refer to this cost function as $C(m,t)$. Write an equation for C in terms of m and t.

★d. What is $C(15,2)$?

79. If a function is defined by the equation $f(x)=x^2+2x+3$, what is the value of $f(10)$?

80. Use the function $f(t)=t^2-3t+5$ to evaluate each expression below.

a. $f(1)=$ b. $f(-2)=$ ★c. $f\left(\frac{1}{3}\right)=$

81. If $f(x)=4x-3$ and $g(x)=1+x^2$, use these functions to write each expression below.

a. $f(x)+g(x)=$ b. $2\cdot f(x)=$ c. $3g(x)=$

82. If $f(x)=4x-3$ and $g(x)=1+x^2$, use these functions to evaluate each expression below.

a. $f(2)+g(3)=$ b. $2\cdot f(3)=$ c. $3g(1)=$

83. If $f(x)=4x-3$ and $g(x)=1+x^2$, use these functions to evaluate each expression below.

a. Since $g(2)=5$, then $f[g(2)]=$ ________.

b. Since $f(-1)=-7$, what is the value of $g[f(-1)]$?

c. What is $g[f(2)]$?

d. Evaluate $f[g(-1)]$.

84. ★If $f(x)=x^2$ and $g(x)=7-x$, then what is the value of $g(f[g(4)])$?

85. Consider the function $f(x)=3+4x+x^2$.

a. Evaluate $4f(0)$.

b. What is the value of $f(-2)$?

★c. For what value of x does $f(x)=0$?

★d. How would you find the expression that represents $f(2M)$?

86. ★If $g(x)=x^2$, then what expression represents $g(M+1)$?

87. Two circus clowns, Emmy and Flo, are launched out of separate canons at the same time. Emmy's canon sits on a platform that is higher than Flo's platform. They fly through the air until they land in a large pool. Their paths through the air are modeled in the graph shown.

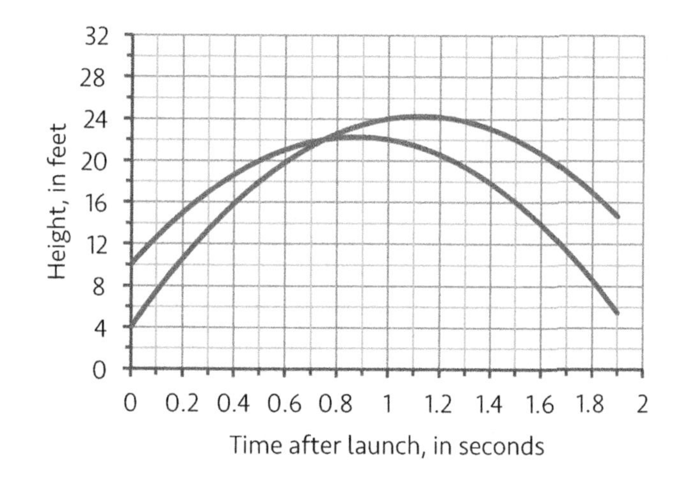

a. Which clown reaches a greater maximum height?

b. Which clown hits the water first?

c. After 1.5 seconds, Flo's height is how much higher than Emmy's?

★d. Which clown traveled farther horizontally during her launch?

88. The Bottle Company produces a refillable drinking bottle. There is a cost associated with producing the bottles, but the company overcomes these costs by selling the bottles at a high enough price. The amount of money that they earn for selling the bottles is called their revenue. The graphs for the revenue, $R(x)$, and costs, $C(x)$, are shown below, where x represents the number of bottles that have been produced and sold in a given month. Assume that every bottle that is produced is also sold.

a. Estimate the value of $C(11)$. Estimate the value of $R(14)$.

b. Estimate how many bottles were produced last year if the total revenue was \$37,500.

c. How many bottles need to be produced to keep the total costs below \$125,000?

89. ★Use the previous scenario to answer each question below.

a. How much does The Bottle Company charge for each bottle?

b. In order for a company to calculate their profit, they must subtract their costs from their revenue. What is The Bottle Company's profit if they produce 11,000 bottles?

c. If you were in charge of this company, would you try to get your workers to increase the production of bottles as much as possible every month? Explain your reasoning.

90. ★A golf tournament is set up to raise money for a local charity. The entry fee is \$80 per golfer. The tournament will award prize money to the top 4 finishers. The best golfer will be awarded the largest amount of money and each additional award will be \$100 less than the previous award.

a. If the total amount of awarded prize money is \$2,000, how much will each of the four best golfers be awarded?

b. Write a function that relates the total amount of awarded prize money to the value of the first place award.

c. If the total prize money that is awarded is 25% of the sum of the entry fees, how many golfers need to sign up for the tournament to make the first place award worth \$800?

91. Consider the two functions defined below.

$K(x)=x^2+x-5$ $\quad\quad N(y)=y-3$

a. What is $K[N(2)]$?

b. What is $N[K(-4)]$?

92. ★In the previous scenario, what is $K[N(y)]$?

93. Use the function $g(x)=2x^3-5x$ to evaluate each expression below.

a. $g(-1)=$

b. $g(-2)=$

★c. $g(\frac{1}{2})=$

94. In the previous scenario, $g(1)=-3$. This can be shown as an ordered pair, (___,___). For each function value that you found in the previous scenario, combine the x and $g(x)$ values to write them as an ordered pair.

Use this page to record important ideas in the previous section or for any other writing that helps you learn the topics in this book.

Section 6
CUMULATIVE REVIEW

95. A plot of land is sold to a developer who splits the land into 4 equally sized sections. If the original plot of land was $2\frac{1}{2}$ acres, how large is each section? Express the size in three forms: as a fraction, as a decimal, and as a percent of an acre. Try to do this without a calculator.

96. Write the first 10 prime numbers. The first prime number in your list should be 2.

97. Rearrange each equation to isolate y. Write your result as a fraction and as a decimal.

a. $\frac{4}{5}y=2$

b. $\frac{1}{4}y-1=-y+2$

98. Rearrange the equation $5x-2y=6$ to isolate y.

99. Without doing any numerical calculations, explain which expression has a larger value.

Expression #1: $2\div\frac{1}{4}$

Expression #2: $2\div\frac{1}{5}$

100. Without a calculator, simplify each expression below.

a. $5\div\frac{1}{6}$

b. $\frac{10}{11}\div 5$

101. Solve each inequality.

a. $3-x>-5$

b. $-2x+4\leq-1+3x$

102. Graph each of the solutions in the previous scenario on a separate number line.

103. Isolate y in each inequality below.

a. $4x+2y\leq18$

b. $-3x-5y>-25$

104. Without graphing, which point is the closest to the x-axis: (3, 2), (−2, −3), (2, 3), or (5, −2)?

105. Is the equation below true if $x = -3$ and $y = -2$?

$$-3x+9y=-8$$

106. If a hiker can travel 9.6 miles in 3 hours, how far will the hiker travel in 4 hours if she maintains the same pace?

Use this page to record important ideas in the previous section or for any other writing that helps you learn the topics in this book.

Section 7
ANSWER KEY

1.	a. "height" is dependent on "age" $\rightarrow$ "height" is the dependent quantity b. "area" is dependent on "radius" $\rightarrow$ "area" is the dependent quantity c. "calories burned" depends on "time" $\rightarrow$ "calories burned" is the dependent quantity
2.	a. $H(a)$ b. $A(r)$ c. $C(t)$
3.	a. dependent: how much sugar dissolves independent: temperature of the water b. dep.: height of the candle indep.: minutes that have passed c. dep.: how many pennies you can stack indep.: number of layers of paper
4.	a. $F=10(0)+3 \rightarrow F=3$ b. $F=10(4)+3 \rightarrow F=43$ c. F is dependent because its value depends on the value of x.
5.	a. \$150 b. The number of people is the independent quantity because the total cost depends on the number of people. c. $C = 75 + 25n$
6.	a. $25+2(10) \rightarrow 45$ in. b. 2 in./day c. The bamboo's initial height was 25 inches when you brought it home.
7.	a. - b. 40 days
8.	a. 225 b. 60 days
9.	a. 185 inches b. 325 inches c. 50 days
10.	Write 80 in the blank. $\rightarrow H(80)$
11.	a. $H(3)=31$ b. $H(11)=47$ c. $H(27)=79$
12.	a. $H=25+2(1) \rightarrow H=27$ inches b. $H=25+2(115) \rightarrow H=255$ inches c. $87=25+2d \rightarrow d=31$ days d. When you buy the bamboo, you have owned it for 0 days and its height is 25 in.
13.	a. $H(13)$ is the height of the bamboo after you have owned it for 13 days. b. The second bamboo plant grows slower (1.9 in./day) than the first plant (2 in./day).

14.	a. 60 miles b. 4 hours c. 70 miles d. 5 hrs
15.	a. 40mph b. 20mph c. 10mph
16.	a. 8 miles b. 45 minutes c. during the first 30 minutes of the ride, the speed was approx. 24 miles per hour
17.	a. $R(5)=8$ b. $n = 7$
18.	a. $y=2(3)+5 \rightarrow y=11$ b. solve $-11 = 2x + 5 \rightarrow x=-8$
19.	a. e b. V or $V(e)$ c. $V(10)=9(10)+2 \rightarrow V(10)=92$ d. solve $-7 = 9e + 2 \rightarrow e=-1$
20.	a. $37 \rightarrow$ simplify: $7-3(-10)$ b. $-14 \rightarrow$ solve: $7-3g=49$
21.	a. 4 b. 1 c. No value. The function is not defined when $n = -10$. d. No value. The function is not defined when $n = 7$.
22.	a. $n = -6$ b. 3 values: $n = -5.5$, 2 and 4 c. $n = -3, -1$ d. No value. The function is not defined when $C(n)=-7$.
23.	a. $f(-6)=(-6)^2-5 \rightarrow 36-5 \rightarrow f(-6)=31$ b. $95=x^2-5 \rightarrow 100=x^2 \rightarrow x=10$ or -10
24.	a. The boy's distance away from home after he has been riding for 12 minutes. b. 12 - 8 = 4 miles. The sister is 4 miles away from her brother at 6:48pm. c. $S(36)=0$, because the sister leaves 36 minutes after her brother, so she has not traveled any distance when $m = 36$.
25.	a. miles per minute b. 15 mph (12 miles in 48 minutes) c. 25 (sister: 40 mph)
26.	decreasing 0 (zero)
27.	The function is not increasing between the x-values of -2 and 4.
28.	Look for a section where the graph is horizontal.
29.	a. $-5<x<1$ b. $4 \leq x<20$
30.	a. $-6<x<-3$ b. $-2<x<0$ and $3<x<5$

31.	a. $3<x<4$ b. $-5<x<-2$
32.	a. strictly increasing b. decreasing then increasing c. switches between decreasing and constant
33.	a. 4 b. 2 c. 1.5
34.	a. −3 and 4 b. 2 c. −7, 5, and 8 d. $g(x)$ never equals −3
35.	$2<x<6$ (between the x-values of 2 and 6)
36.	3 intervals: $-7<x<-1$, $-1<x<2$, $6<x<8$
37.	least: −7
38.	greatest: 8
39.	least: 2 greatest: 6
40.	$-2<n<3$
41.	2 intervals: $-6<n<-2$ and $3<n<4$
42.	least: −6 greatest: 4
43.	least: −5 greatest: 5
44.	a. $(-2)^2-2(-2)-3\to4+4-3\to T(-2)=5$ b. $0=h^2-2h-3\to0=(h-3)(h+1)$ $\to h=3$ or -1
45.	a. The constant portion of each graph occurs when Sam waits for the bus. In the morning, he waits for the bus after walking for 20 minutes. In the afternoon, he waits for the bus after walking 5 minutes. b. Sam waits for the bus 5 minutes longer in the afternoon than he did in the morning. c. 6 mph (2 miles in 20 min.) d. about 30 mph (slightly less than 8 miles in 15 min.)
46.	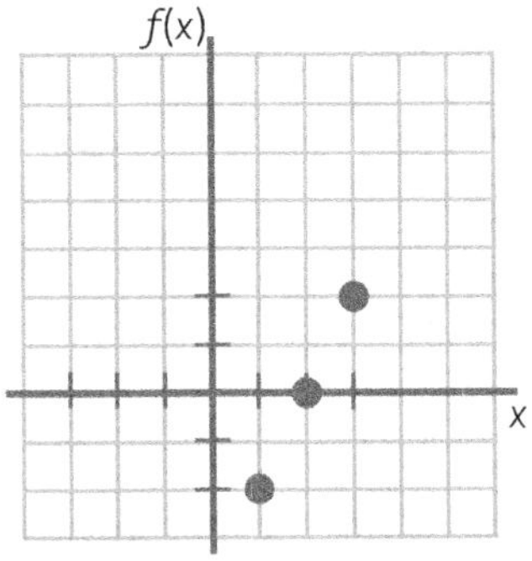 a to c.
47.	a. 4 b. 8
48.	$f(x)=2x-4$
49.	a. least x-value: −6 greatest x-value: 7 b. least y-value: 2 greatest y-value: 6
50.	a. $1\le x\le7$ b. $4\le y\le16$
51.	a. $1\le x\le6$ b. $x\ge15$
52.	a. $20\le y\le80$ b. $y\ge21$
53.	a. all real numbers, or using an inequality, it can be written as $-\infty<x<\infty$

	b. $6\le x\le11$
54.	a. $y\ge2$ b. 20
55.	a. The line continues in both directions without ending. b. all real numbers (all of the numbers that can be shown on the number line.
56.	all real numbers (all of the numbers that can be shown on the number line
57.	Infinity is an idea that allows us to show that something can go on without ending. The value of "infinity" also represents a number that is beyond what you can get to by counting.
58.	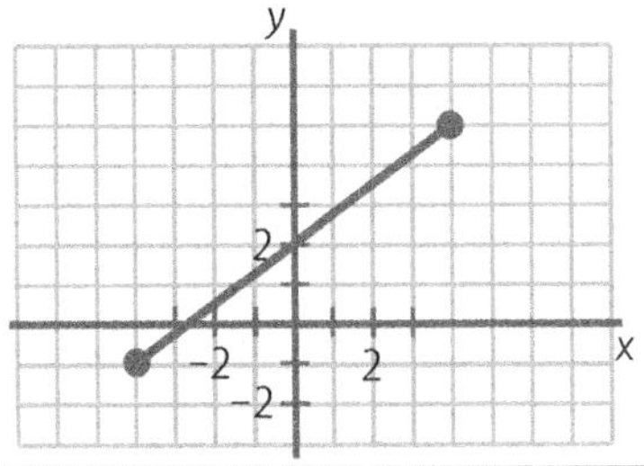
59.	Yes, there are an uncountable number of decimal values between 0 and 1.
60.	segment
61.	a. d: {Penny, Nickel, Dime, Quarter} b. r: {1, 2, 3, 4}
62.	a. domain: {−5, −4, 0} b. range: {−2, 1, 3, 6}
63.	range: 1, 4, 7, 10, 13
64.	$\{-6, 0, 14, 50\}$
65.	a. −6 and 4 b. $-6\le x\le4$ c. 1 and 5 d. $1\le y\le5$
66.	$y = 1$; $y=\frac{2}{3}x+3$; $y = -2x + 11$
67.	a. $-6\le x\le-3$ b. $-3\le x\le3$ c. $3\le x\le4$
68.	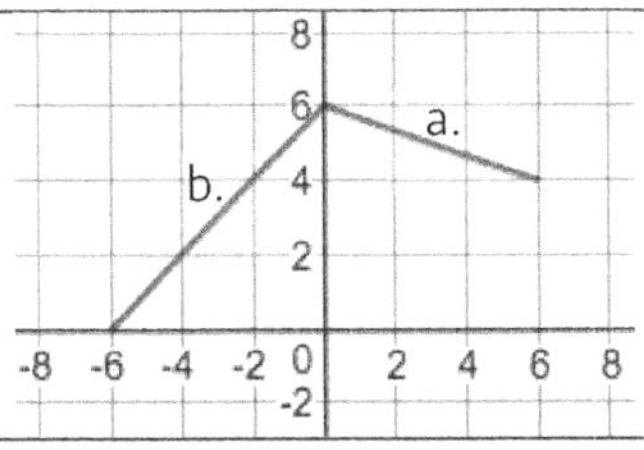
69.	a. 5 b. 4+6 = 10
70.	a. −6 and 6 b. $-6\le x\le6$ c. 0 to 6 d. $0\le y\le6$
71.	a. 10 miles b. 2.5 miles c. 3 t-values
72.	Time is the independent quantity, because the distance depends on the time.
73.	The horizontal axis, or the x-axis, is typically used to represent the independent quantity.
74.	a. $T(0)$ is the temperature at midnight

	b. $T(15) - T(3) = 98° - 66° = 32°$, which is the difference between the highest and lowest temperatures shown in the graph c. $h = 9$ and 21 (9am and 9pm) d. between $h = 0$ and 3 and between $h = 15$ and 21.
75.	a. 0 and 21 b. 66 and 98 c. $0 \le h \le 21$ d. $66 \le T \le 98$
76.	a. \$430 b. 42 hours c. employees earn \$10 per hour until they work 40 hours d. employees earn \$15/hr for every hour above 40 e. $10(40) + 15(11) = \$565$
77.	60 hours
78.	a. \$8.70 b. 7 min. c. $C(m,t) = 2.50 + 2.00m + 0.40t$ d. \$33.30
79.	$f(10) = (10)^2 + 2(10) + 3 = 123$
80.	a. 3 b. 15 c. $4\frac{1}{9}$
81.	a. $(4x-3)+(1+x^2) \to x^2+4x-2$ b. $8x - 6$ c. $3 + 3x^2$
82.	a. $5+10 \to 15$ b. $2 \cdot 9 \to 18$ c. $3 \cdot 2 \to 6$
83.	a. 17 b. 50 c. 26 d. 5
84.	-2; $g(4)=3 \to f(3)=9 \to g(9)=-2$
85.	a. $4 \cdot 3 \to 12$ b. $3+4(-2)+(-2)^2 \to -1$ c. $0=3+4x+x^2 \to x = -1, -3$ d. In the function, replace x with $2M$.
86.	$(M+1)^2 \to M^2+2M+1$
87.	a. Flo b. Emmy c. 6 ft (22 ft − 16 ft) d. There is not enough information to answer this question. The graphs show time, but they do not show horizontal distance.
88.	a. \$75,000; \$125,000 b. around 4,000 c. approximately less than 15,500
89.	a. about \$9 b. approx. \$25,000 c. No, after 18,000 bottles are produced the costs become higher than the revenue so the company would have negative profits (they would be losing money). Thus, the production needs to be kept lower than 18,000.
90.	a. 350, 450, 550, 650 b. FP + (FP−100) + (FP−200) + (FP−300)=PM $\to$ 4FP − 600 = PM (FP is 1st place prize, PM = prize money) c. 130 golfers

91.	a. $K[N(2)]=-5$ First, find $N(2)$. $N(2)=2-3 \to -1$ Next, find $K(-1)$. $K(-1)=(-1)^2+(-1)-5 \to -5$ b. $N[K(-4)]=4$ $K(-4)=(-4)^2+(-4)-5 \to 7$ $N(7)=7-3 \to 4$
92.	$K(y-3)=(y-3)^2+(y-3)-5$ $y^2-6y+9+y-3-5 \to y^2-5y+1$
93.	a. 3 b. −6 c. $-\frac{9}{4}$
94.	$g(1)=-3 \to (1,-3)$ a. $(-1,3)$ b. $(-2,-6)$ c. $\left(\frac{1}{2},-\frac{9}{4}\right)$
95.	fraction: $\frac{5}{8}$ decimal: 0.625 percent: 62.5%
96.	2, 3, 5, 7, 11, 13, 17, 19, 23, 29
97.	a. $\frac{5}{4}\cdot\frac{4}{5}y=2\cdot\frac{5}{4} \to y=\frac{10}{4} \to y=\frac{5}{2}$ or 2.5 b. $\frac{5}{4}y=3 \to \frac{4}{5}\cdot\frac{5}{4}y=3\cdot\frac{4}{5} \to y=\frac{12}{5}$ or 2.4
98.	$-2y=6-5x \to y=\frac{6-5x}{-2} \to y=-3+\frac{5}{2}x$
99.	Expressions 2 is larger because $\frac{1}{5}$ is smaller than $\frac{1}{4}$. If you divide 2 into smaller portions, there will be more portions overall. For example, if you divide 2 cookies into sections that are each one-fifth of a cookie, there will be more sections than if you divide 2 cookies into sections that are each one-fourth of a cookie.
100.	a. $5 \div \frac{1}{6} \to 5 \cdot \frac{6}{1} = 30$ b. $\frac{10}{11} \div 5 \to \frac{10}{11}\cdot\frac{1}{5}=\frac{2}{11}$
101.	a. $-x>-8 \to \frac{-x}{-1}>\frac{-8}{-1} \to x<8$ b. $-5x \le -5 \to \frac{-5}{-5}x \le \frac{-5}{-5} \to x \ge 1$
102.	a. 8 b. 1
103.	a. $2y \le 18-4x \to y \le 9-2x \to y \le -2x+9$

	b. $-5y > -25 + 3x \rightarrow y < 5 - \frac{3}{5}x \rightarrow y < -\frac{3}{5}x + 5$
104.	Two points are closest to the x-axis. The points (3, 2) and (5, −2) are both 2 units from the x-axis.
105.	No; $-3(-3) + 9(-2) \rightarrow 9 - 18 \rightarrow -9 \neq -8$
106.	9.6 miles in 3 hours is a rate of 3.2 miles per 1 hour. In 4 hours, that would be 12.8 miles.

HOMEWORK & EXTRA PRACTICE SCENARIOS

As you complete scenarios in this part of the book, you will practice what you learned in the guided discovery sections. You will develop a greater proficiency with the vocabulary, symbols and concepts presented in this book. Practice will improve your ability to retain these ideas and skills over longer periods of time.

There is an Answer Key at the end of this part of the book. Check the Answer Key after every scenario to ensure that you are accurately practicing what you have learned. If you struggle to complete any scenarios, try to find someone who can guide you through them.

CONTENTS

Section 1

COMPARING INDEPENDENT AND DEPENDENT QUANTITIES

1. A function shows how one quantity affects another quantity. Since there are usually two quantities involved in a function, it is common to call one of them dependent, because its value depends on the value of the other quantity. Identify the dependent quantity in each statement below.

   a. The price of a bag of rice is a function of the weight of the bag.

   b. The amount of time needed to paint lines on an athletic field is a function of the total length of the lines that need to be painted.

   c. The water pressure felt by a scuba diver is a function of (is impacted by) how deep the diver is below the surface.

2. Since the circumference of a circle, C, is a function of the length of its radius, r, you can quickly refer to the circumference function as C of r, or $C(r)$. In each statement below, name the function using the two variables.

   a. The price of a bag of rice, P, is a function of the weight of the bag, w. The price function is ______.

   b. The amount of time needed to paint the lines, T, is a function of the total length of the lines, l. The time function is ______.

   c. The water pressure on a scuba diver, P, is a function of how deep the diver is below the surface, d. The pressure function is ______.

3. Read the description of each experiment. Identify the dependent variable and the independent variable in the experiment.

   a. Roll a car down a ramp. Start the car at different heights on the ramp and see how far the car travels along the floor after it gets to the bottom of the ramp.

   b. Go to a frozen yogurt store. Each time a customer buys a bowl of frozen yogurt, write down the weight of the frozen yogurt mixture and the price of the mixture. Analyze how the price of the mixture changes with the weight of the mixture.

   c. Plant a seed. Once the plant starts to grow, record its height every day and see how it's height changes over time.

Section 2

REPRESENTING A FUNCTION WITH AN EQUATION OR A GRAPH

There are many ways to represent a function, or to show how one quantity affects another quantity. Two common ways to display a function are through an equation or a graph.

4. Consider the function represented by the equation $H = t^2 + t + 5$. In this equation, H is a function of t because the value of H depends on the value of t.

   a. If $t = 0$, what is the value of H?

   b. If $t = -5$, then $H =$ _____ .

   c. For the function $H = t^2 + t + 5$, which variable is the independent quantity?

5. Consider the function represented by the graph. The graph shows how the cost of the apples depends on the total weight of the apples.

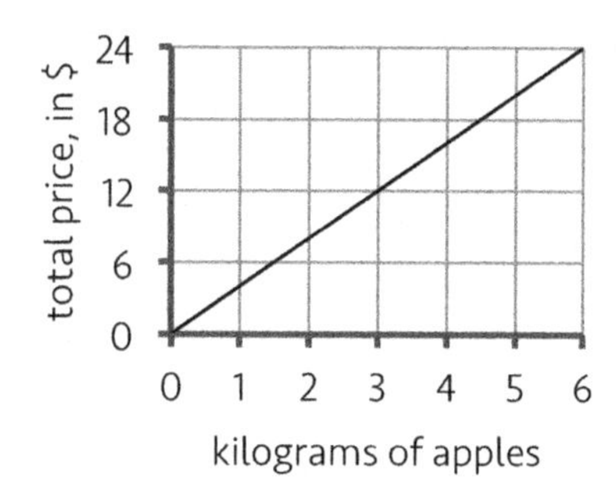

   a. What is the cost of 4.5 kilograms of apples?

   b. How many apples did someone buy if they paid $12?

   c. How many kilograms of apples can you buy for $28?

   ★d. Write an equation that shows the total price, P, if someone buys k kilograms of apples.

6. If you mail a large envelope, using the United States Postal Service First-Class pricing, there is a function that determines the price. The price function is $P = 0.60 + 0.20w$, where P is the price, in dollars, to send an envelope that weighs w total ounces.

   a. What is the price to mail an envelope that weighs 10 ounces?

   b. Use the function to find the price you pay if you mail an envelope that weighs 0 ounces. Does it make sense to find this price?

   c. How much does the price increase for each additional ounce of weight?

7. The function in the previous scenario can be shown in a graph, with the two quantities shown on the two axes.

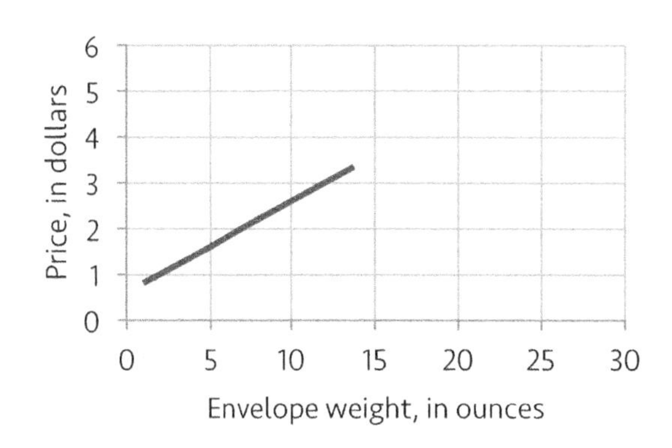

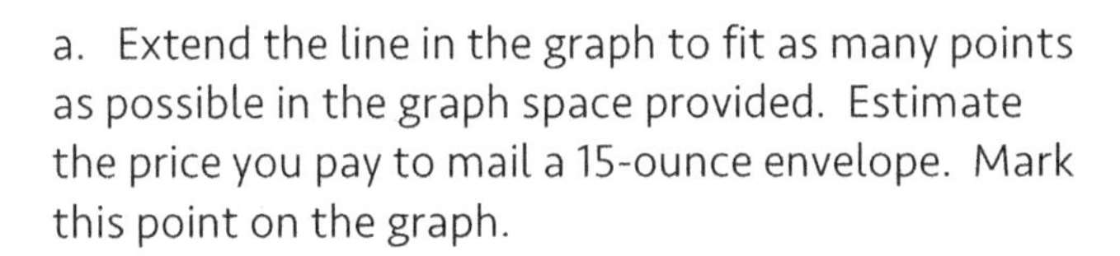

a. Extend the line in the graph to fit as many points as possible in the graph space provided. Estimate the price you pay to mail a 15-ounce envelope. Mark this point on the graph.

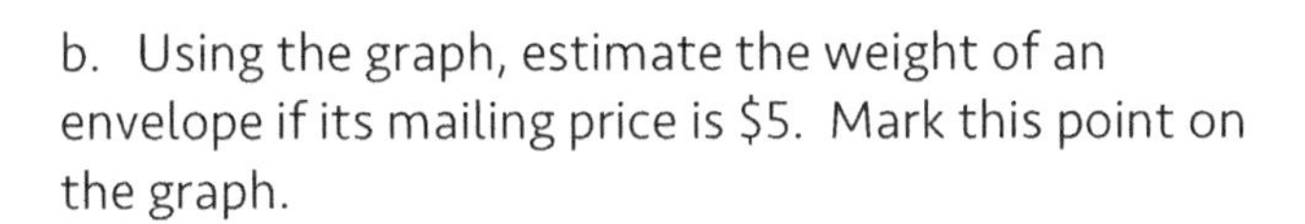

b. Using the graph, estimate the weight of an envelope if its mailing price is \$5. Mark this point on the graph.

c. How can you check the accuracy of your estimate in part b.?

8. The function in the previous scenario can also be shown in a table, with the columns labeled to easily identify the two quantities.

weight, w	Price, P
7	\$2.00
12	\$3.00
17	\$4.00
25	\$5.60
	\$6.80
39	

a. Fill in the missing values in the table.

★b. Using the table, if you send 2 envelopes and the larger one weighs 20 more ounces than the smaller one, what will be the price difference between the two envelopes?

9. In the previous scenario, the mailing price, P, is a function of the weight of the envelope, w. A more concise way to state this is that P is a function of w. A brief way to refer to this function is to call it "P of w," which uses two letters to identify the two quantities related. In the previous scenario, you found that the price for sending 10-ounce envelope is \$2.60. Using this more concise language, you could state that P of 10 is \$2.60.

a. What is P of 17?

b. What is P of 40?

10. In the previous scenario, what is w if the value of "P of w" is \$10?

11. The expression P of w is written as $P(w)$ by mathematicians. Using this notation, the question "What is P of 17?" can be rewritten as "What is $P(17)$?". Rewrite each statement below in more concise function notation. The first one is started for you.

a. The price for a 10-ounce envelope is \$2.60. $\rightarrow P(10)=$ _____

b. To send a 17-ounce envelope, the price is \$4.00. $\rightarrow$ ______________

c. It costs \$5.60 to send a 25-ounce envelope. $\rightarrow$ ______________

12. As a reminder, the price function was defined earlier by the equation $P = 0.60 + 0.20w$.

    a. Using this equation, what is $P(1)$?

    b. What is w if $P(w) = \$5$?

    c. What is $P(-1)$?

13. The graph of the function $V(k)$ is shown.

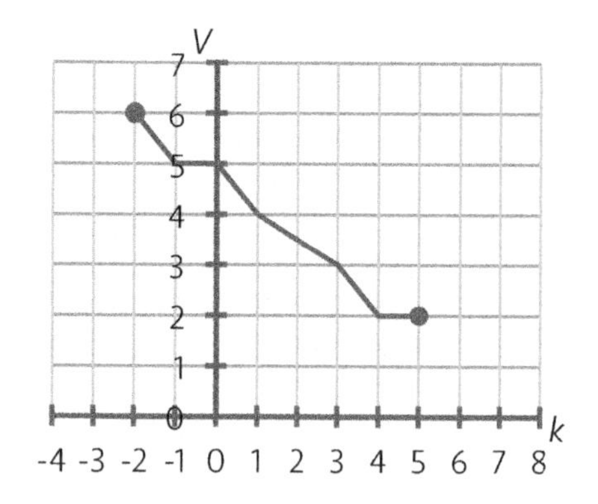

    a. What is the value of $V(-2)$?

    b. What is the value of k if $V(k) = 4$?

As a reminder, a function shows the relationship between two quantities. If you think about these quantities as x's and y's, then the first question in the previous scenario involves finding a y-value when you are given a specific x-value. In the second question, you are given a y-value and you are asked to find the x-value that is paired with it.

14. Consider a function given by the equation $y = -7x + 11$.

    a. What is the value of y if $x = -5$?

    b. What is the value of x if $y = 60$?

15. Consider the function given by the equation $T(n) = 10 - 3n$.

    a. You may be more familiar with having the variables x and y in your equations. For the function $T(n) = 10 - 3n$, which variable is taking on the role of "x" in this equation?

    b. Which variable is "y" in the function $T(n) = 10 - 3n$?

    c. The notation $T(n)$ shows that T gets its value from n. What is the value of $T(5)$?

    d. What is the value of n, if the value of $T(n) = 10$?

16. A function is defined by the equation $H(y)=\dfrac{y-5}{6}$. Fill in each blank below.

a. $H(-7)=$ ________ b. $H($_____$)=11$

17. Use the graph to find the value of each expression listed below. Estimate if necessary.

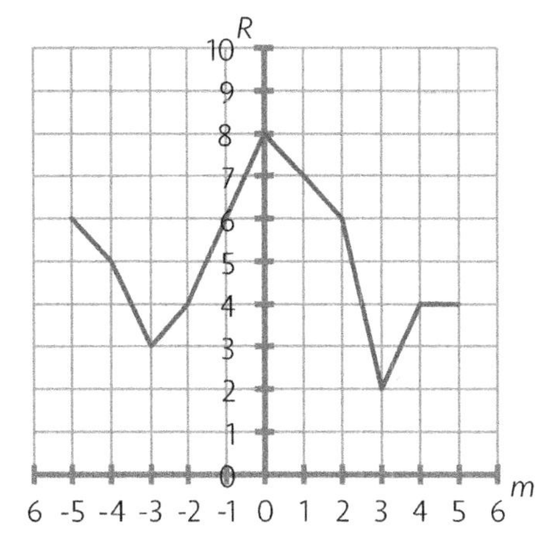

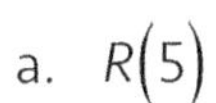

a. $R(5)$

b. $R(-2)$

c. $R(-7)$

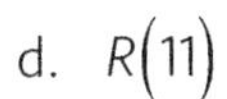

d. $R(11)$

18. Use the graph in the previous scenario to find each value below. Estimate if necessary.

a. What is m if $R(m)=8$?

b. What is m if $R(m)=6$?

c. Fill in the blank.

$R($____$)=3$

d. Fill in the blank.

$R($____$)=0$

19. The function $f(x)$ is defined by the equation $f(x)=2x^2+4x+3$. Fill in each blank below.

a. What is $f(-5)$?

b. What is x if $f(x)=9$?

20. Two brothers, Andy and Nolan, own separate roofing companies. Last summer, they worked for 10 weeks and they kept track of their total earnings each week. If Andy's earnings are assigned the variable A, Nolan's earnings are assigned the variable N, and time is represented by t, for total weeks worked, then their earnings can be represented by separate functions, $A(t)$ and $N(t)$. The two functions are shown in the graph below.

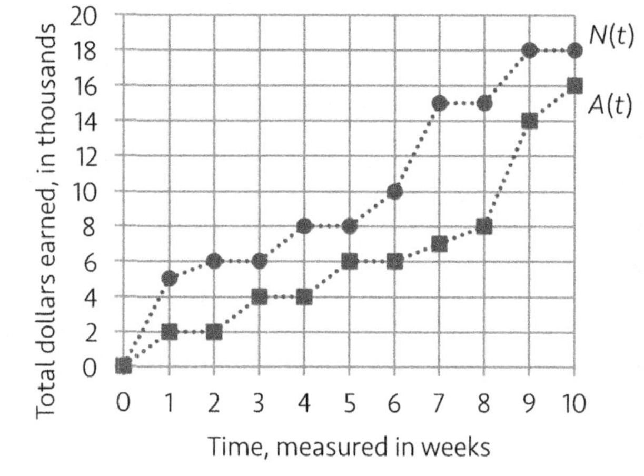

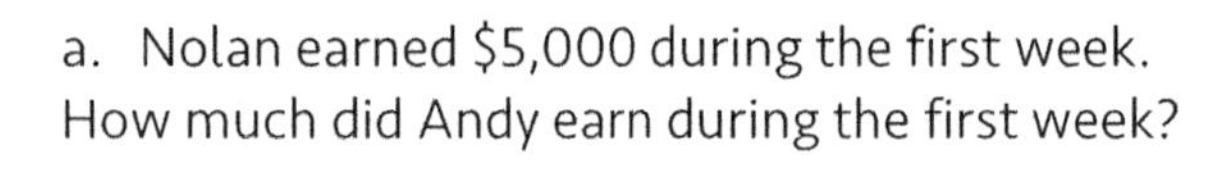
a. Nolan earned \$5,000 during the first week. How much did Andy earn during the first week?

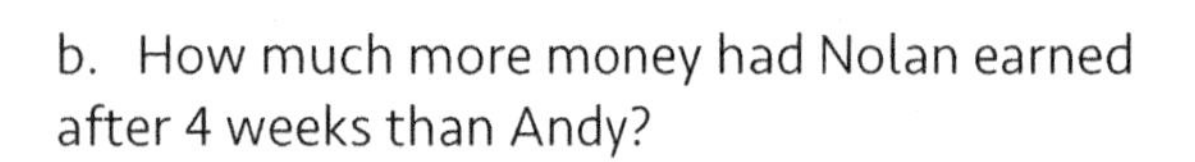
b. How much more money had Nolan earned after 4 weeks than Andy?

c. After how many weeks had Nolan earned \$8,000 more than Andy?

d. Estimate the value of $N(8)-A(8)$.

21. Use the graph in the previous scenario to work through each of the following questions.

a. For what value of t does $A(t)=\$7{,}000$?

b. Which brother had the best single week of earnings?

c. Which brother had the greatest number of weeks without any earnings?

22. ★In the previous scenario, who had the greatest average increase in earnings after week 5 (the second half of the graph)? Compute his average earnings per week during this 5-week time period.

23. A farm in Georgia trains dogs. The graph shows how the daily dog population on the farm changed over time. Define $P(y)$ as the function that shows the relationship between the dog population, P, and the number of years, y, that have passed since 2000.

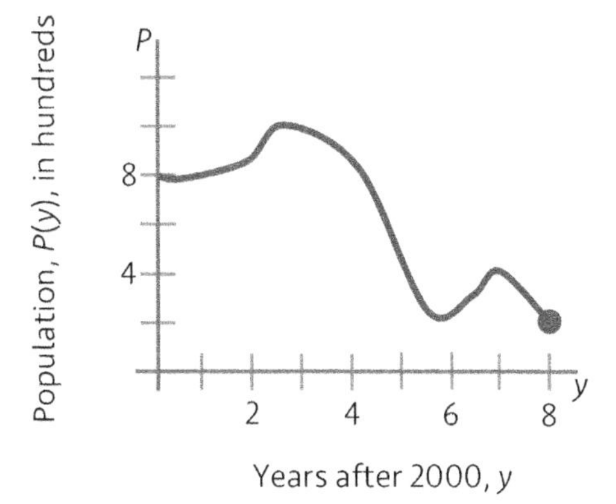

a. Estimate the value of $P(8)$.

b. Estimate the value of y for which $P(y) = 400$.

24. ★Refer to the graph in the previous scenario.

a. During which year (from January 1 to December 31) did the population decrease by the greatest amount?

b. During which year (from January 1 to December 31) did the population increase by the greatest amount?

25. The business in the previous scenario grew slowly each year until another dog-training business opened up nearby and became very popular.

a. What year did the farm start to experience a significant decline in the number of dogs that they kept on the farm?

b. Estimate the total number of years that the dog population experienced a steady decline before it finally started increasing again.

★c. What was the average rate at which the population declined each year from the beginning of 2003 until the beginning of 2005?

Section 3

INCREASING, DECREASING OR CONSTANT

26. Try to fill in the blanks. When a function has a positive slope, the function is described as increasing. When a function has a negative slope, it is described as ________________. When the slope of a function is ___, it is described as constant.

27. A bucket is filled with 10 cups of water, but a small crack in the bottom lets water leak out at a constant rate. After 5 minutes, the bucket is completely empty.

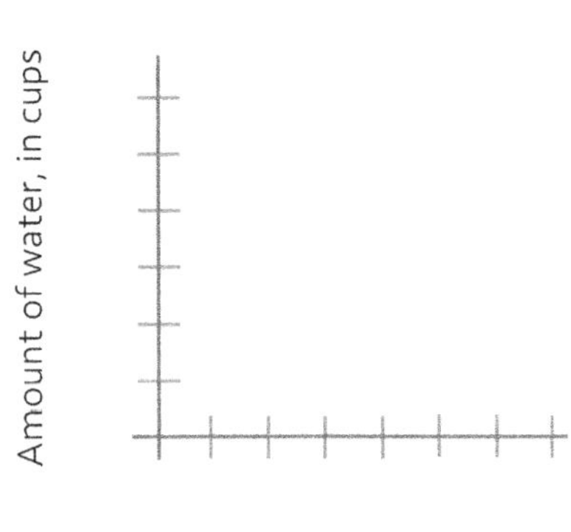

a. Create a graph that shows how the amount of water in the bucket changed from the moment the bucket was filled to the moment the bucket was empty.

b. The graph represents a function that is strictly ________________.

c. Over what interval is the graph defined?

28. ★In the previous scenario, at what rate did the water drain out of the cup?

29. Explain why the function shown is NOT strictly decreasing.

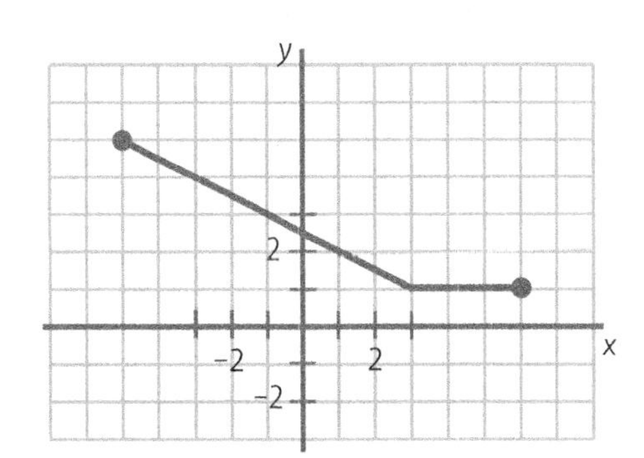

30. The previous function is not increasing in the section between (3, 1) and (6, 1). In this interval, the function is considered to be constant. The word "constant" is used in this case because the function's y-values are not changing as the x-values increase. The y-values are constant. What can you look for in the graph of a function to find a section where the function is constant?

31. The previous function is constant between 2 points, (3, 1) and (6, 1). Another way to state this is to focus on the x-values and say that the function is constant between the x-values of 3 and 6. A more concise way to show this is with interval notation: $3 < x < 6$ (the points that have x-values between 3 and 6). Write each statement below using interval notation.

    a. between the x-values of −10 and 14, but not including −10 or 14

    b. between the x-values of −1 and 7, including −1 and including 7

32. Identify the interval(s) on which the following function is/are constant. Write the interval(s) as an inequality.

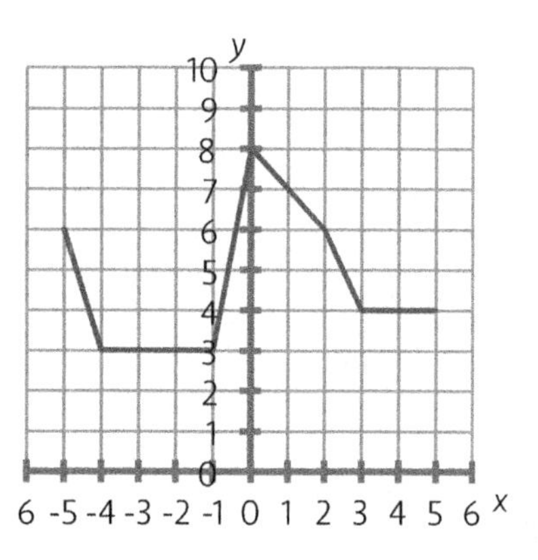

33. Identify the interval(s) on which the function in the previous scenario is increasing.

34. Identify the interval on which each function is constant. Write the interval as an inequality.

a.

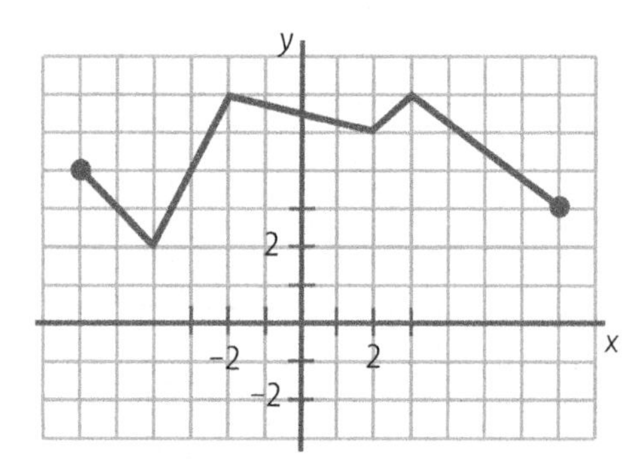

b.

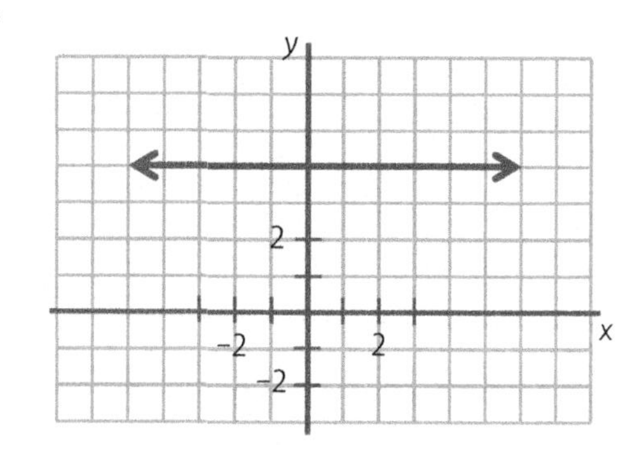

35. Identify the interval(s) on which each function in the previous scenario is decreasing.

36. After a summer soccer practice, players walk over to the water cooler and take turns refilling their water bottles. The water cooler initially contains 8 liters of water and each player takes the same amount of water before letting the next player fill up. After 4 minutes, the cooler is completely empty and 8 players have filled up their water bottles.

a. During this 4-minute time period, is the amount of water in the cooler strictly decreasing?

b. How much water was put into each water bottle?

37. ★Draw a graph that shows how the amount of water in the cooler changes as the players fill up their bottles in the previous scenario. Your graph should be a rough estimate, but try to make it as accurate as possible.

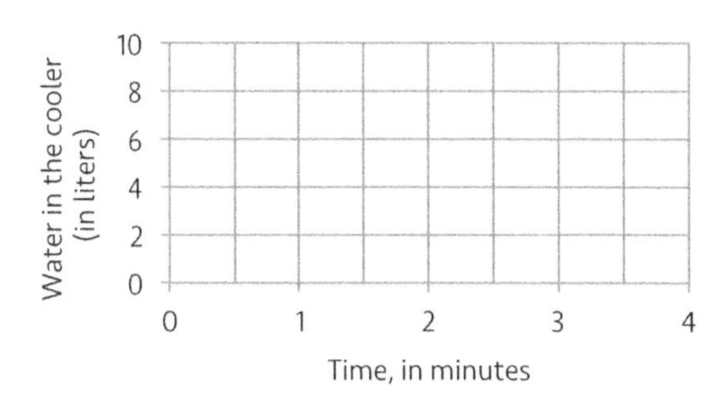

38. Describe each function below using the words increasing, decreasing, or constant.

a.

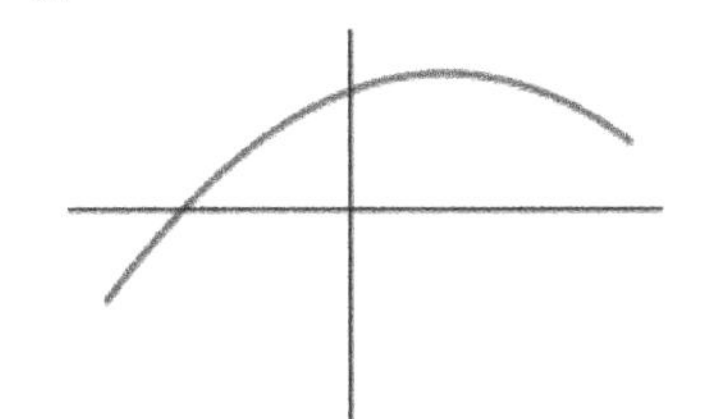

b.

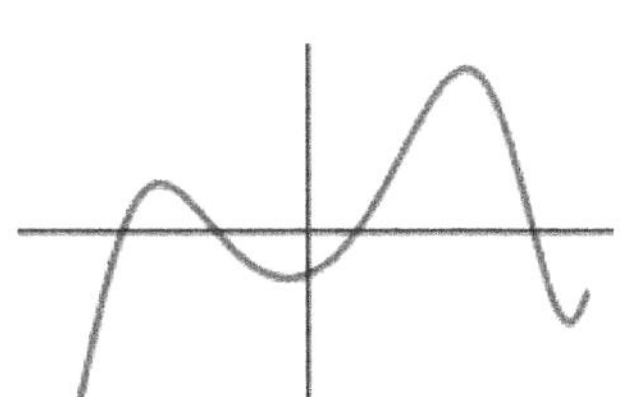

c.

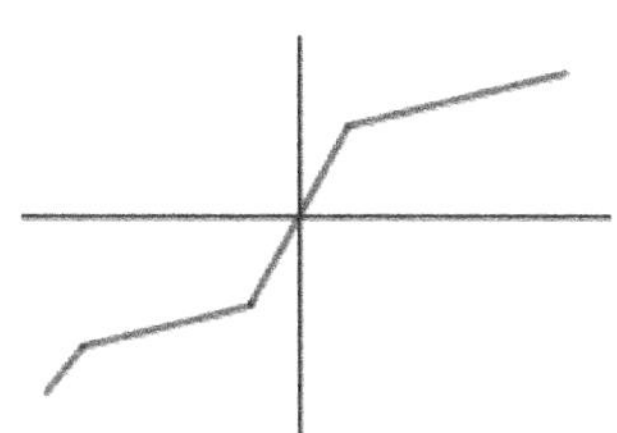

39. Consider a function that is called $g(x)$.

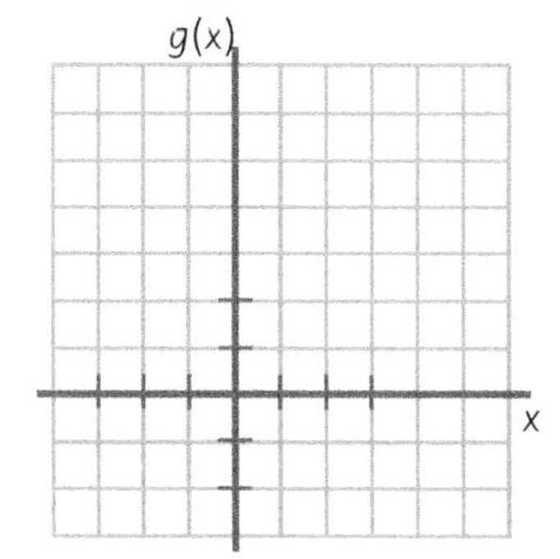

a. The value of $g(1)$ is 3. Show this as a point in the graph to the right.

b. The value of $g(2)$ is 2. Show this in the graph.

c. $g(4)=0$. Show this in the graph.

d. If $g(x)$ is a linear function, what is the value of $g(10)$?

40. In the previous scenario, $g(x)$ is identified as a linear function. Write the equation for $g(x)$, in Slope-Intercept Form.

41. The graph of a function $g(x)$ is shown. It only contains the points between the endpoints, as shown. Determine the value of each expression listed below.

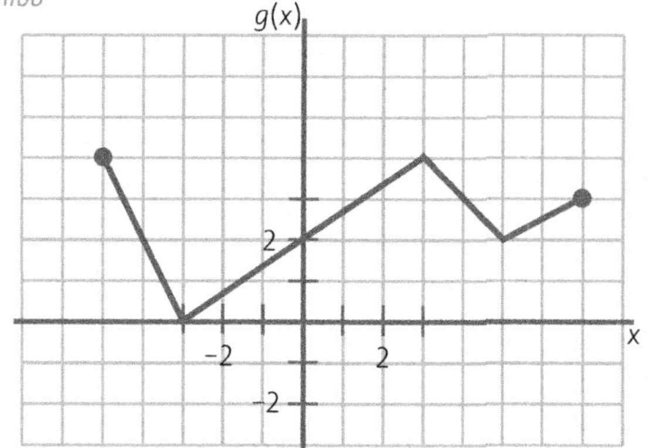

a. $g(0)$

b. $g(-3)$

c. $g(6)$

42. Use the graph in the previous scenario to find the value of each expression listed below.

a. x when $g(x)=4$

b. x when $g(x)=5$

c. x when $g(x)=2$

d. x when $g(x)=1$

43. For what values of x is the function increasing in the previous scenario?

44. Refer to the graph in the previous scenario to answer the following questions.

a. What are the least and greatest x-values shown in the function?

b. What are the lowest and highest y-values shown in the function?

45. Two functions are shown in the same graph.

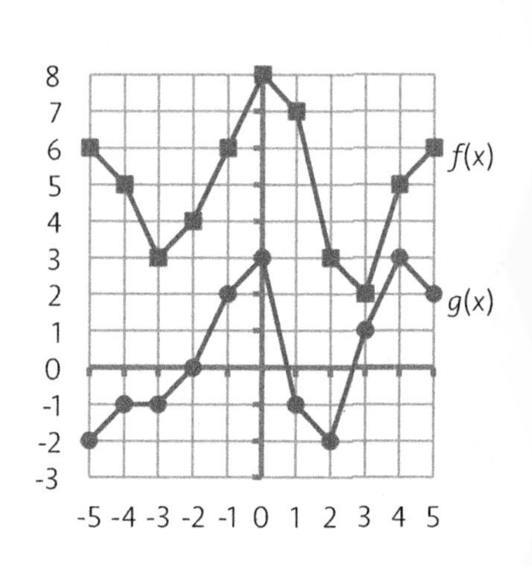

a. What is $f(0)+g(0)$?

b. Which is greater, $g(4)$ or $f(3)$?

c. Over what interval is $g(x)$ constant?

Section 4

THE DOMAIN AND RANGE OF A FUNCTION

At this point, you are ready to learn two new mathematical words: domain and range. The **domain** of a function is the set of all of the x-values of the function. The **range** of a function is the set of all of the y-values of the function.

46. A function is shown in the graph.

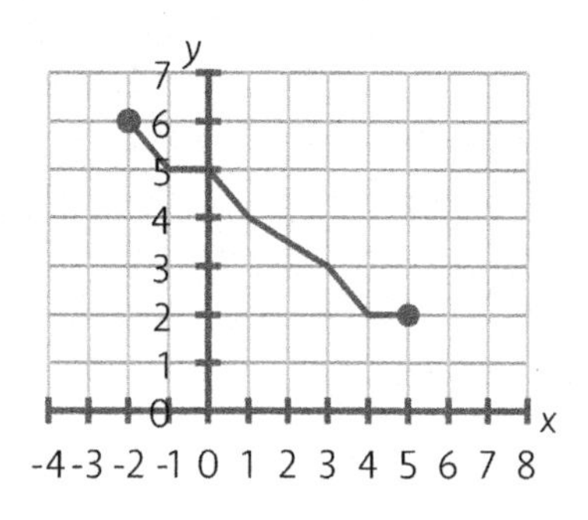

a. What is the domain of the function?

b. What is the range of the function?

47. Identify the domain of each function shown below.

a.

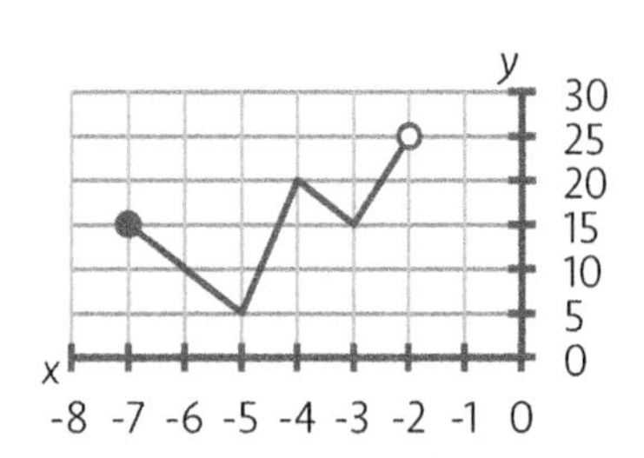

b.

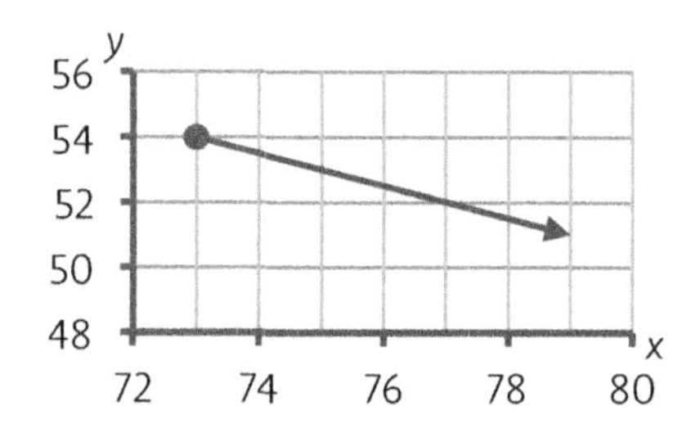

48. Identify the range of each function in the previous scenario.

49. Identify the domain of each function shown below.

a.

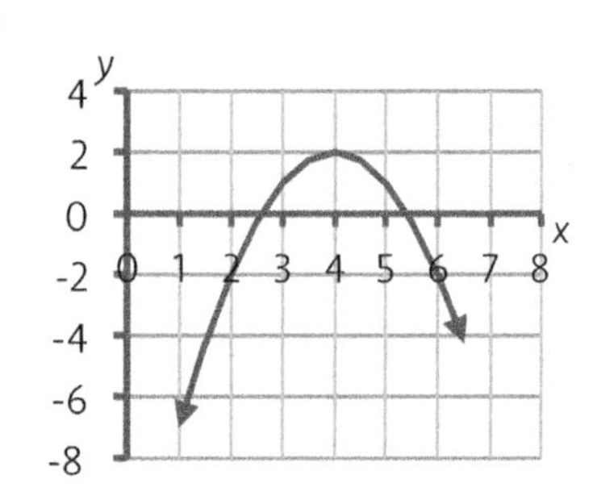

b.

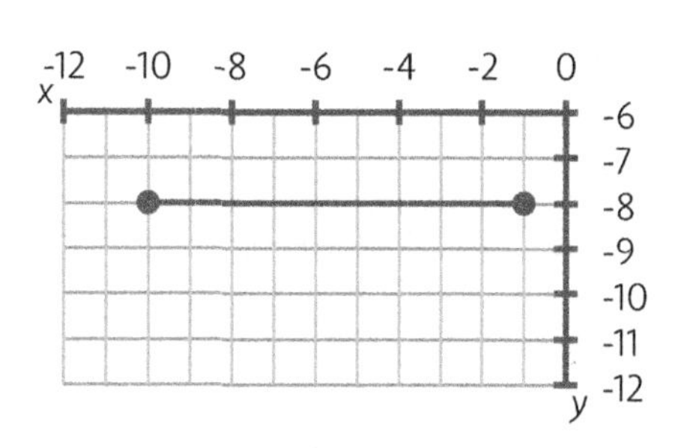

50. Identify the range of each function in the previous scenario.

51. Identify the domain and range of the function shown.

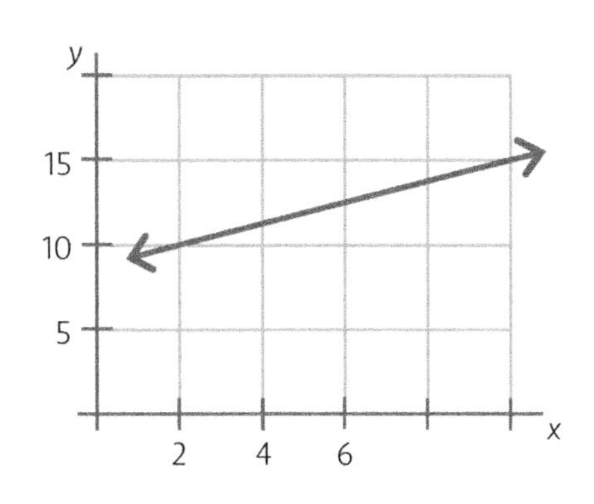

52. Consider the empty grid shown to the right.

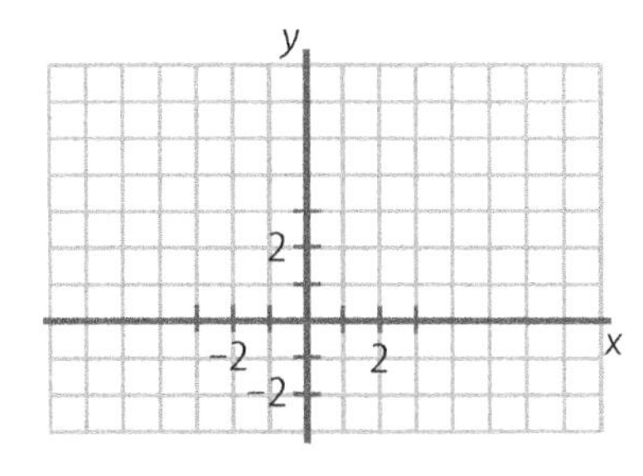

a. Graph the line given by the equation $y = -x + 4$.

b. What is the domain of the line $y = -x + 4$ if you graph all of the points that are part of this line?

53. If you restrict the domain in the previous scenario, and only graph the part of the line that has *x*-values from −3 to 6, the line will become smaller. Graph the line if its domain is only the *x*-values from −3 to 6.

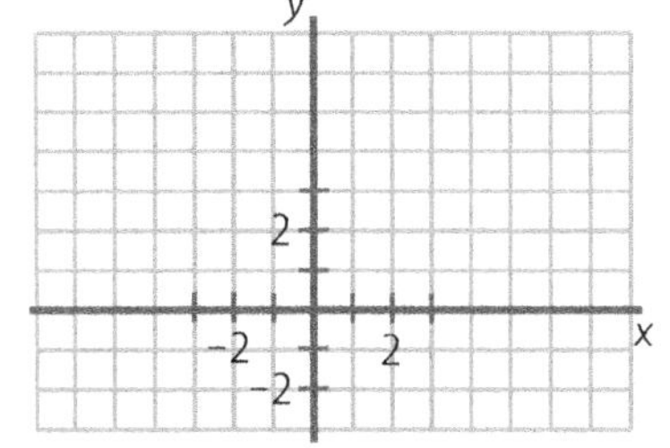

54. Graph the function $y = \frac{1}{2}x + 1$, but restrict the domain to the *x*-values from −2 to infinity. In other words, graph the part of the line that starts at $x = -2$ and continues slanting upward to the right without ending.

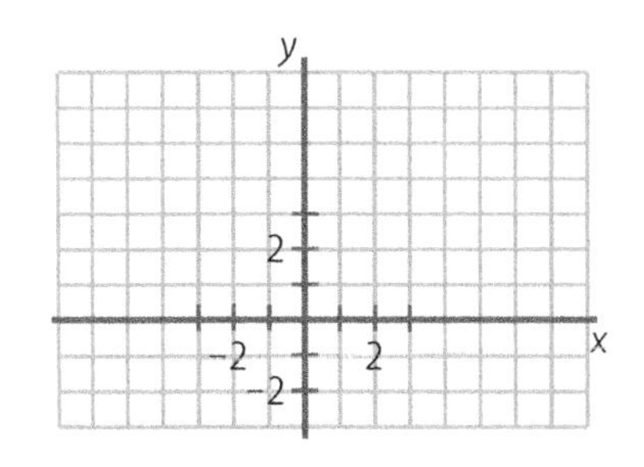

55. Using a compound inequality, you can write the domain of the previous function as $-2 \leq x < \infty$. Using words instead of symbols, this inequality states that *x* can be any value greater than or equal to −2, but it must be less than infinity. Why do you think the inequality does not state that *x* can be less than or equal to infinity?

56. Graph the function $y = x + 5$, but restrict the domain to the *x*-values that are less than or equal to 0. In other words, graph the part of the line that starts at $x = 0$ and continues slanting downward to the left without ending.

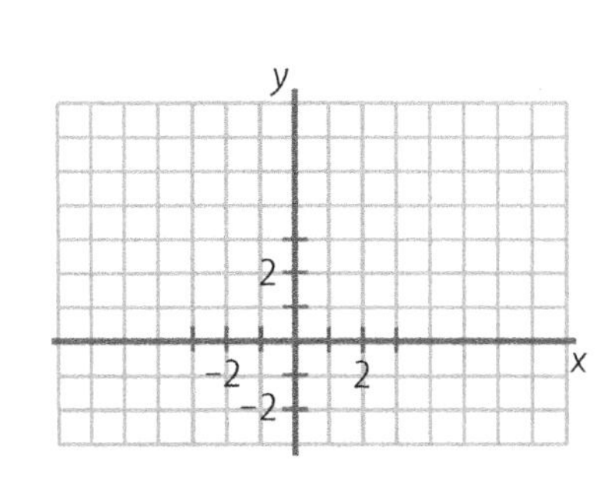

57. Write the domain of the previous function, using a compound inequality.

58. A Geometry concept is worth mentioning here. By definition, a line extends without ending in both directions. If one of the arrows is replaced with an endpoint and the line now only extends in one direction without ending, the line is now called a ___________.

59. Consider a scenario that does not have an infinite domain. At the beginning of the school year, students attend a meeting in the gym. Students are sorted into specific seating sections in the bleachers based on their grade level. For this sorting function, the domain is the student's grade level (these are the inputs) and the range is the section where they sit (the outputs). Suppose the sorting function works as shown.

Grade Level	Freshman	Sophomore	Junior	Senior
Seating Section	Orange	Yellow	Blue	Green

a. What is the domain of the function?

b. What is the range of the function?

60. Suppose there is a simple function that is only defined for the six ordered pairs shown. State the domain and range of this function.

$$\{(5,3), (-1, -1), (0, 0), (2, -1), (7, 0), (-4, 6)\}$$

61. A function is defined by the equation $f(x) = -x - 7$. Suppose the domain is restricted to only five values: 0, 1, 2, 3, and 4. Since these values are the restricted domain, they are the only x-values that you can use in the equation. Calculate the value of $f(x)$ for each of the given x-values and combine these values to show the restricted range of this function.

62. Another function is shown below and it is assigned a restricted domain. Use the given domain to find the range of each function.

function: $g(x) = x^2 - 3$ domain: $\{-2, -1, 0, 1, 2\}$ range:

63. Consider the given function. It only contains the points between the endpoints, as shown.

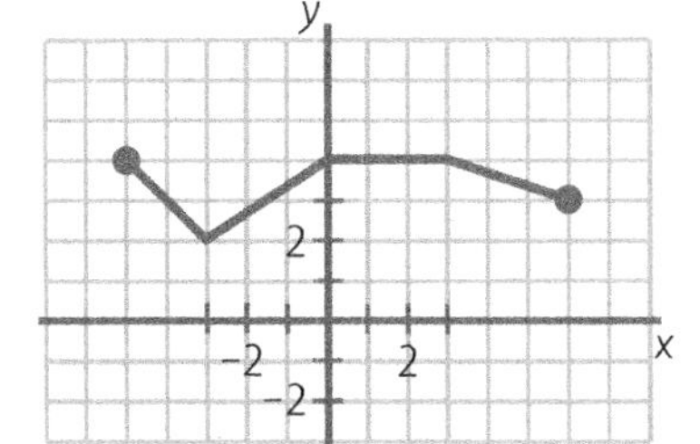

a. What are the lowest and highest x-values shown in the function?

b. The domain of the function is the x-values between _____ and _____. As an inequality, the domain can be written as _____ $\le x \le$ _____ .

c. What are the least and greatest y-values shown in the function?

d. The range of the function is the y-values between _____ and _____. As an inequality, the range can be written as _____ $\le y \le$ _____ .

64. There is not a single equation that will produce the graph in the previous scenario. It has been formed by taking segments from 4 different lines. Write the 4 equations that would represent the 4 lines.

65. If you graphed each of the 4 lines that you identified in the previous scenario, the lines would extend in both directions without ending. In the previous scenario, however, only a small portion of each line is shown because they have restricted domains. Identify the restricted domain for each line.

a. The restricted domain for the line given by the equation $y = -x - 1$ is _____ $\le x \le$ _____ .

b. The restricted domain for the line $y = \frac{2}{3}x + 4$ is ________________.

c. The restricted domain for the line $y = 4$ is ________________.

d. The restricted domain for the line $y = -\frac{1}{3}x + 5$ is ________________.

66. Graph the function that is described below. It is made by graphing a small portion of 2 separate lines.

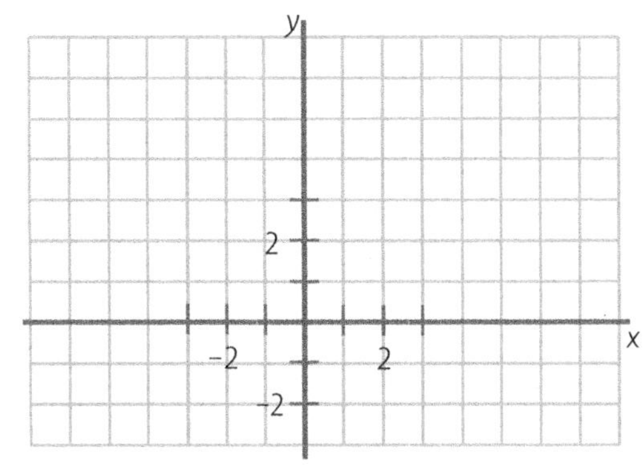

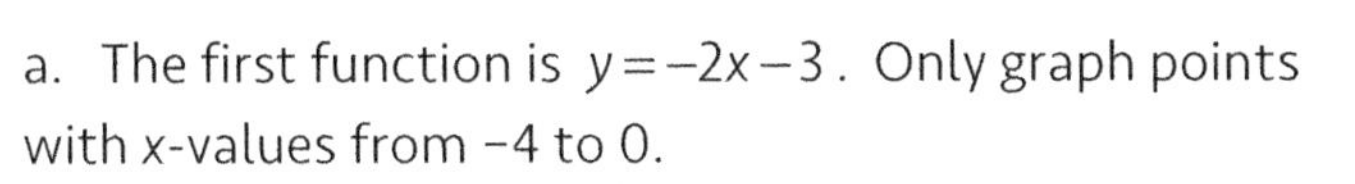

a. The first function is $y = -2x - 3$. Only graph points with x-values from −4 to 0.

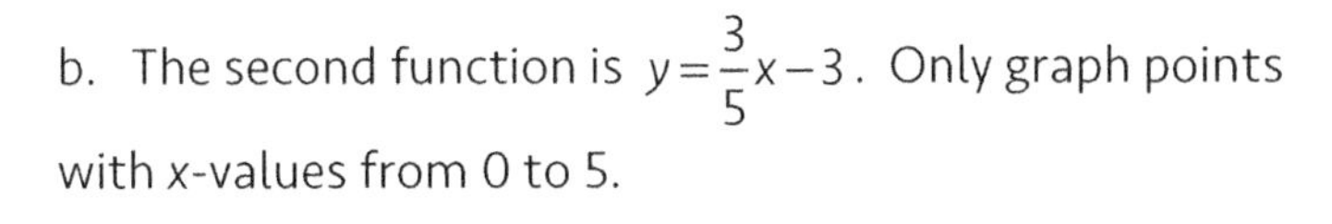

b. The second function is $y = \frac{3}{5}x - 3$. Only graph points with x-values from 0 to 5.

c. If the entire function is given the name $H(x)$, what is the value of $H(-3) + H(0)$?

67. Refer to the graph in the previous scenario to answer the following questions.

a. What are the least and greatest x-values shown in the function?

b. The domain of the function is the x-values between ____ and ____. As an inequality, the domain can be written as ____ $\leq x \leq$ ____ .

c. What are the lowest and highest y-values shown in the function?

d. The range of the function is the y-values between ____ and ____. As an inequality, the range can be written as ____ $\leq y \leq$ ____ .

68. Consider the function shown in the graph.

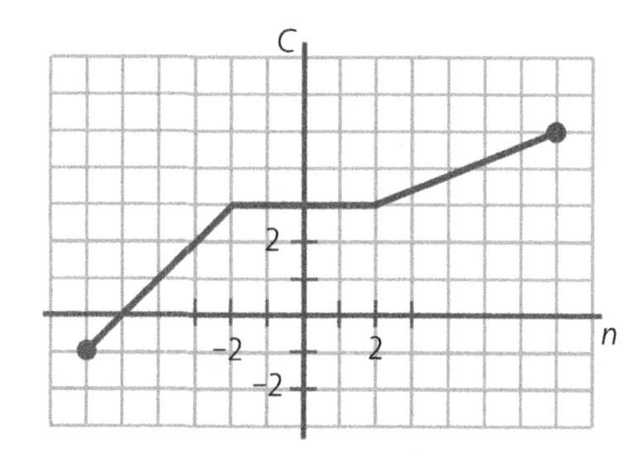

a. Identify the domain of the function.

b. Identify the range of the function.

69. Consider the function shown in the graph.

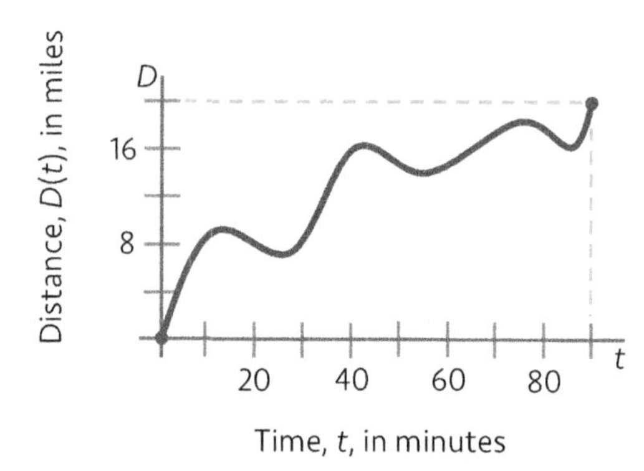

a. Identify the domain of the function.

b. Identify the range of the function.

70. In the previous scenario, for how many t-values does $D(t) = 8$?

71. The previous scenario compares two quantities. Which one is the dependent quantity?

72. Using what you have seen so far, when a function is plotted in a graph, which axis is used to represent the dependent quantity?

73. Identify the independent variable in each graph shown.

a.

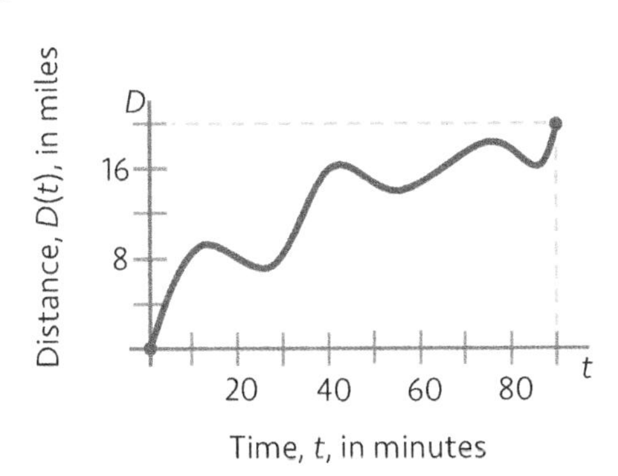

b.

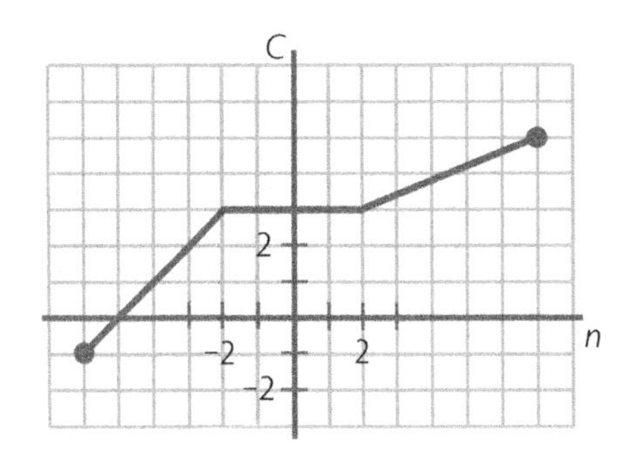

74. Consider the graph. It shows how the height of a rocket changes as it flies through the air. The rocket reaches a maximum height of 150 feet and it lands on the ground exactly 7 seconds after it is launched.

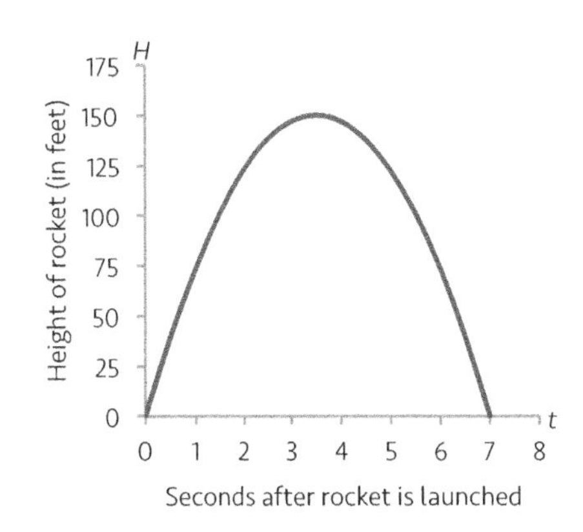

a. Estimate the value of $H(1.5)$.

b. Explain why the range of the graph is not correctly labeled if it is written as $0 < H < 150$. Write the range in its correct form.

c. Identify the domain of the graph.

d. In this scenario, which quantity is the dependent quantity?

75. For the function $f(x)$, the ordered pair $(\_\_\_\_, \_\_\_\_)$ is represented by the statement $f(6)=2$.

76. Write the ordered pair that is represented by the statement $g(-5)=0$.

Section 5

MORE SCENARIOS THAT INVOLVE FUNCTIONS

77. In 2014, the largest Ferris Wheel in the world was the High Roller in Las Vegas, NV. If you are unfamiliar with Ferris Wheels, they are upright, rotating circles. Riders get on near the ground and then travel around the outer edge of the rotating circle until they achieve a complete loop. This loop may continue multiple times. The graph below displays how a rider's height changes as a function of the amount of time that the rider spends on the High Roller.

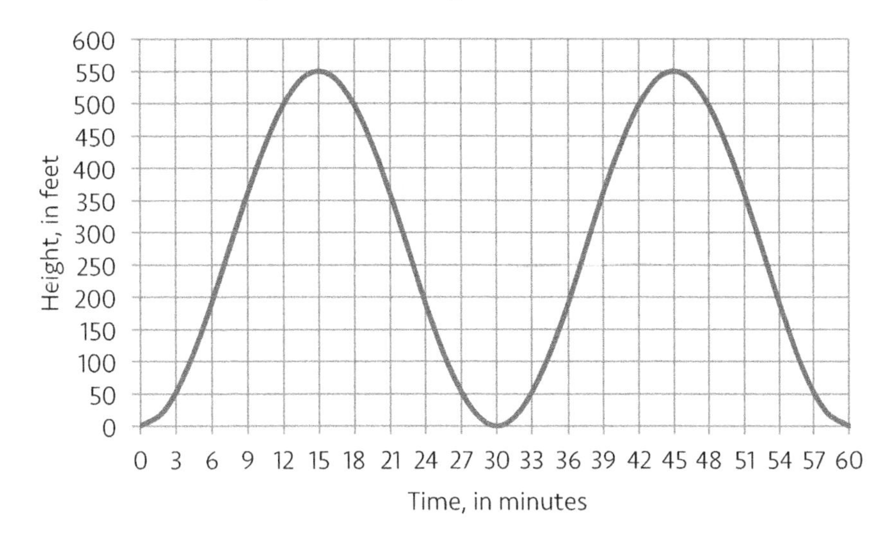

a. How much time does it take to complete one full rotation around the High Roller?

b. What is the maximum height achieved by a rider on the High Roller?

78. Use the graph in the previous scenario to answer the following questions.

a. Estimate the rate at which a rider's height changes between the 6th and 9th minute?

★b. Estimate the rate at which a rider's height changes between the 21st and 24th minute?

79. Refer to the graph in the previous scenario to answer the following questions.

a. What is the domain of the function, if the function is only what is shown in the graph? Write the domain as an inequality.

b. Write the range of the function as an inequality.

80. Suppose a taxi in Los Angeles charges an initial fee of \$3.75 plus an extra \$2.25 per mile. The taxi also considers the impact of traffic and charges \$0.50 for every minute that the taxi is stopped in traffic. How much would a taxi charge for an 8-mile trip that included 5 minutes of time stopped in traffic?

81. In the previous scenario, the taxi's cost function (call it C) depends on two quantities. Let m represent the miles traveled and t represent the time stopped in traffic. Since C is dependent on two quantities, you can refer to this cost function as $C(m,t)$.

    a. The equation for C in terms of m and t can be written as follows (fill in the blanks):

    $C(m,t) =$ ______ + ______m + ______t

    b. What is $C(5,2)$?

    c. If $C(20,9) = \$53.25$, what does that represent in the context of this scenario?

82. ★Refer to the previous scenario to answer the questions below.

    a. If $C(m,12) = \$32.25$, what is the value of m?

    b. If $C(15,t) = \$39.00$, what is the value of t?

83. A function is defined by the equation $g(t) = -2t^2 - t$. What is the value of $g(-2)$?

84. Use the functions $f(x) = -2x^2 + 1$ and $g(x) = 3 - x^3$ to evaluate each expression below.

    a. $f(2) =$

    b. $g(0) =$

    c. $f(-3) =$

    d. $g(-3) =$

    e. $f\left(\frac{1}{2}\right) =$

    f. $g(-5) =$

85. ★If $R(t)=5-2t$ and $S(t)=-t^2+t$, use these functions to evaluate each expression below.

a. What is $R\left[S(-4)\right]$?

b. Evaluate $S\left[R(4)\right]$.

86. Consider the function $N(k)=9-k^2$.

a. Find $N(0)$.

b. What is the value of $N(-3)$?

★c. For what value of k does $N(k)=0$?

★d. How would you find the expression that represents $N(x)$?

87. ★If $N(k)=2k^2+2k$, then what expression represents $N(x+1)$?

88. One afternoon in science class, your teacher allows you to shoot a rocket into the air. The rocket is launched from a platform on the ground. The graph shows how the rocket's height changes after it is launched.

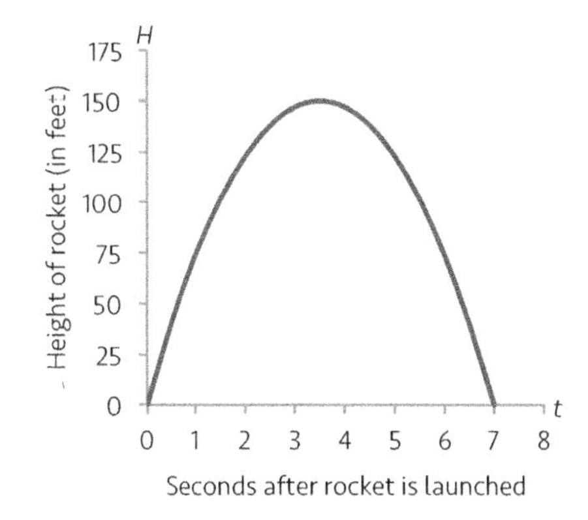

a. Identify the domain of the function.

b. Identify the range of the function.

c. Estimate the value of $H(1)$.

d. Why does $H(7)$ equal 0?

89. Ellen likes to run races that cover short distances, while Anna likes to run really long races, so Ellen challenges Anna to a 1-mile race, since that seems like a good compromise. Details from their race are shown in the graph.

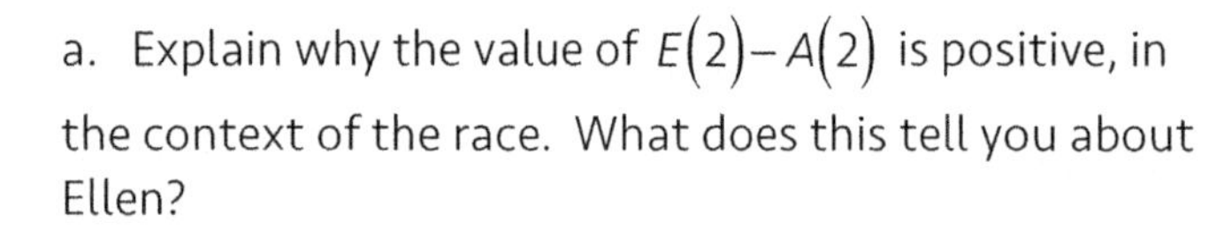

a. Explain why the value of $E(2)-A(2)$ is positive, in the context of the race. What does this tell you about Ellen?

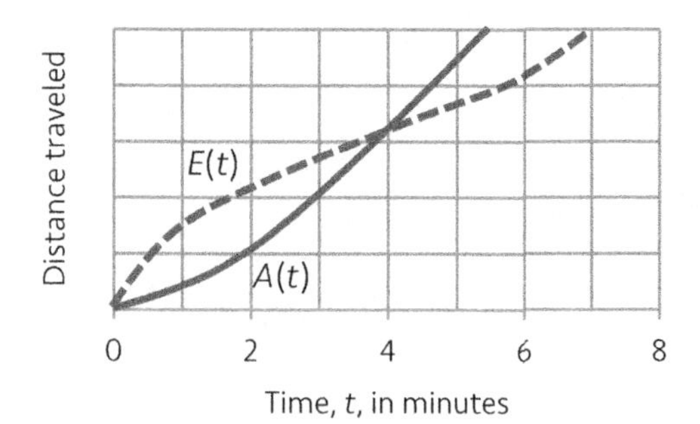

b. Who is running faster after 3 minutes?

c. Explain why the value of $E(5)-A(5)$ is negative, in the context of the race.

90. Describe what happened during the part of the race that is shown in the previous graph.

91. ★Part of the function, $f(x)$, is shown in the graph. The values of this function repeat every 5 x-values. Finish drawing the graph to fit as many values as you can in the space provided.

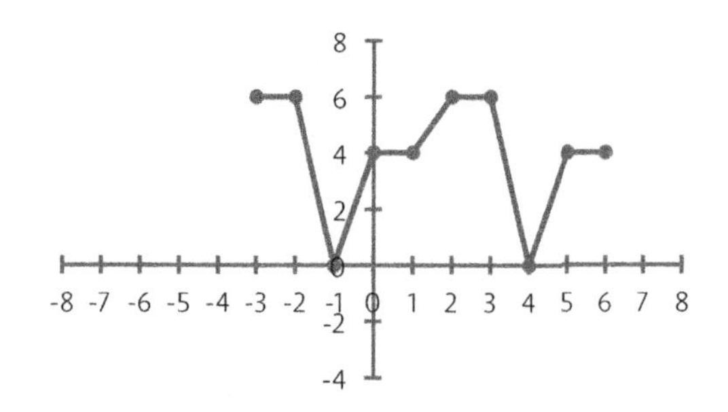

92. ★Refer to the graph in the previous scenario to answer the questions below.

a. What is the value of $f(10)$?

b. What is the value of $f(34)$?

93. Consider the two functions defined below.

$V(t)=12-t^2$ $\qquad$ $R(w)=-2w+5$

a. What is $V[R(4)]$?

b. What is $R[V(-2)]$?

Section 6
BOOK 1 REVIEW

94. What is the ordered pair represented by the statement $f(-4)=7$?

95. Use the graph to find the value of each expression listed below. Estimate if necessary.

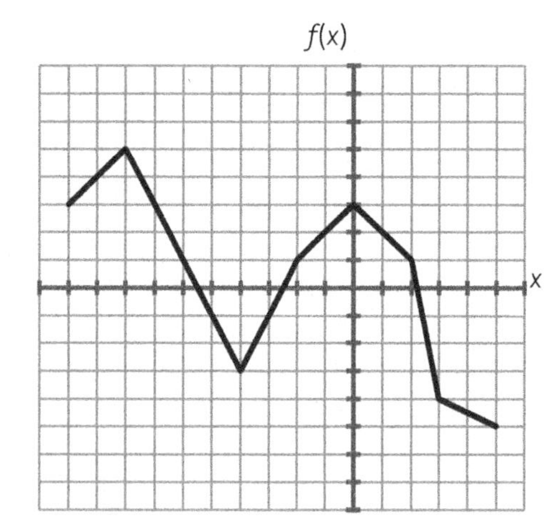

a. $f(4)$

b. x when $f(x) = 1$

c. $f(-8)$

d. x when $f(x) = -4$

96. Given: $h(x)=5x$ and $g(x)=\sqrt{x}$.

a. Find $h(-3)$.

b. What is $h\left[g(64)\right]$?

97. Three functions are shown. Use the functions to find the value of each expression listed below.

$f(x)=\frac{3}{5}x-2$ $\quad g(x)=-4x$ $\quad h(x)=x^2-1$

a. $f(15)$

b. $g(-4)$

c. $h(-3)+g\left(\frac{1}{4}\right)$

98. Three functions are shown. Use the functions to find the value of each expression listed below.

$f(v)=7-\frac{1}{5}v$ $\qquad g(t)=6t$ $\qquad h(x)=x^2+x-2$

a. $f(-5)\cdot h(3)$ $\qquad$ b. $h[f(50)]$ $\qquad$ c. $g(y-2)$

99. $G(x)$ is a function. If $G(-10)=0$, does this represent the function's x- intercept or its G-intercept? How do you know?

100. The domain of a function is $-4 \le x \le 6$. What does this domain tell you about the graph of the function?

101. What is the difference between the domain and the range of a function?

102. Briefly sketch a function that has the domain and range listed below.

Domain: $-4 \le x \le 6$

Range: $-2 \le y \le 5$

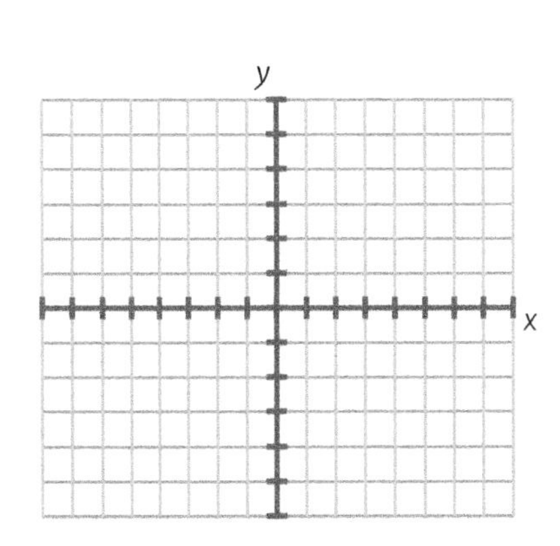

103. Use the graph for the following questions.

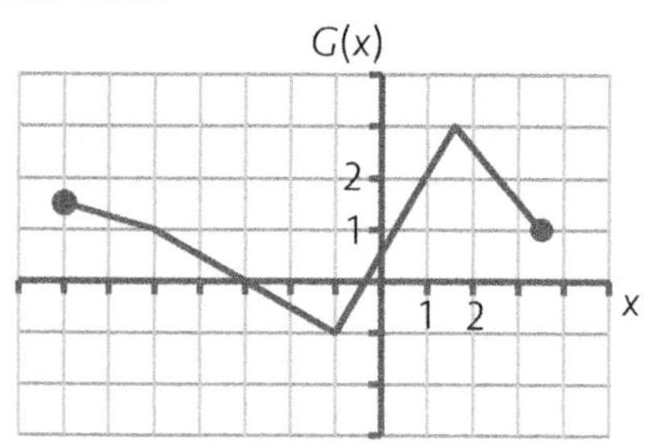

a. Explain in words how you would find $G(-1)$ using the graph.

b. Explain in words how you would find $G(x)=2$ using the graph.

c. What is the range of this function?

d. What is the domain of this function?

104. Identify the interval on which the following function is decreasing. Show the interval using an inequality like ____ $< x <$ ____.

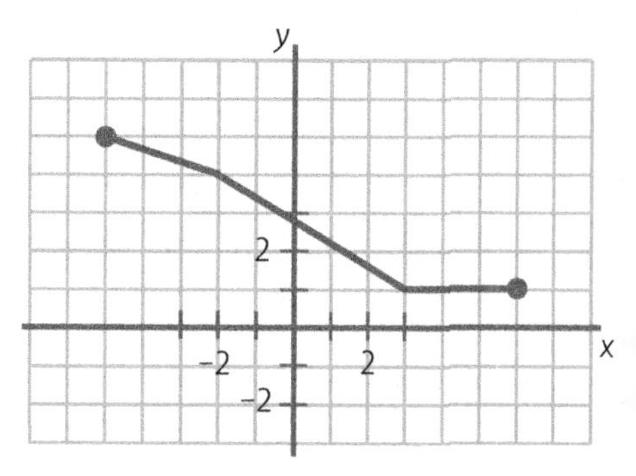

105. The previous function is not decreasing on the interval $3<x<6$ (the points that have x-values between 3 and 6). In this interval, the function is considered to be constant. Identify the interval on which the following functions are constant. If possible, write the interval as an inequality.

a.

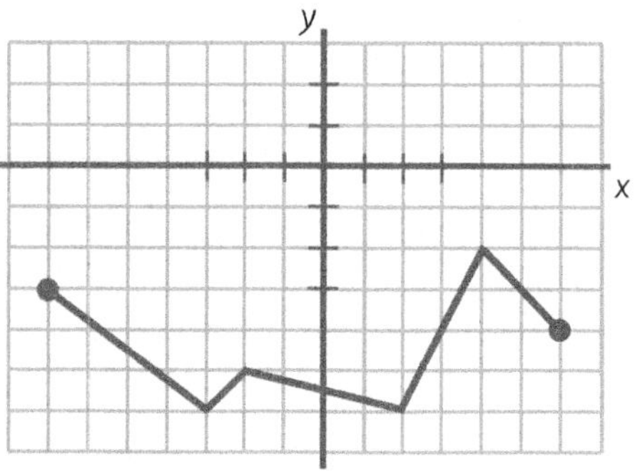

b.

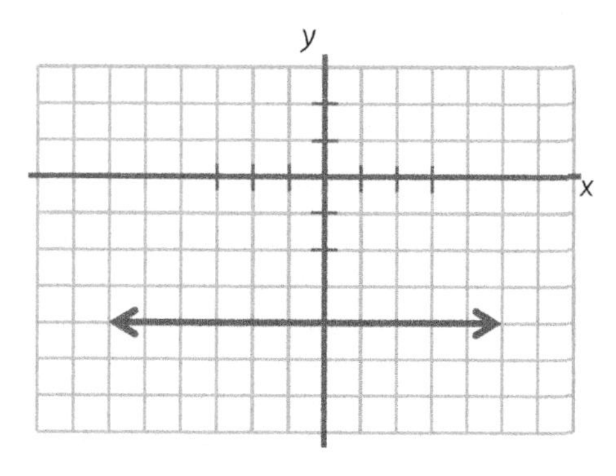

106. Identify the interval on which the graph in the next scenario is increasing.

107. Consider the graph. It shows how the height of a ball changes as it travels through the air after it is thrown.

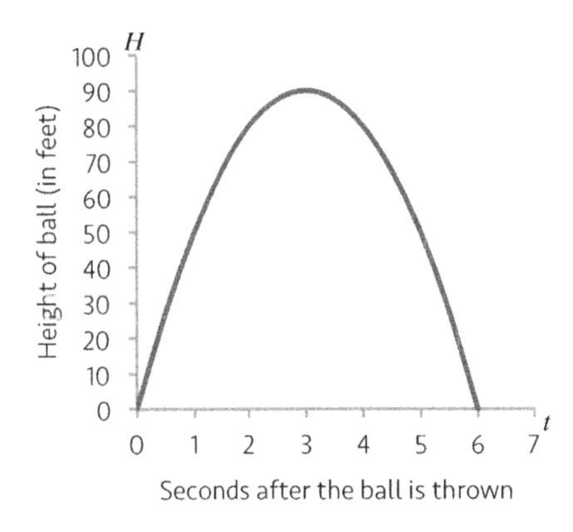

a. Estimate the value of $H(1)$.

b. When does $H(t)=80$?

108. Identify the domain of the graph in the previous scenario.

109. In the previous scenario, which quantity is the dependent quantity? How do you determine which quantity is dependent?

Section 7

ALGEBRA 1 SERIES BOOKS 1–7 REVIEW

110. Simplify the following expressions.

a. $4 \cdot 2 + 6 \cdot 3$

b. $10 - 5(2+1)$

c. $-|9-8|-6$

111. Simplify the following expressions.

a. $5-(-1)+(-2)-6$

b. $\dfrac{(2-6)^2+3^3}{1-2}$

c. $-45+15-(-10)-5+(-25)+50$

112. Evaluate the following expressions for the given variables:

a. x^2-5 when $x=4$

b. x^2-3x+6 when $x=5$

113. Evaluate the following expressions for the given variables:

a. $-|x|+3y$ when $x=-2$ and $y=-6$

b. x^2y-3w when $x=2$, $y=-1$ and $w=4$

c. $\dfrac{xy-y^2}{2x}$ when $x=3$ and $y=-2$

114. What is the value of x that makes each equation true?

a. $x+\frac{1}{8}=\frac{1}{2}$

b. $3x+2x=30$

c. $\frac{-x}{3}=17$

d. $3x-7=11$

115. Solve each equation.

a. $\frac{1}{4}x-\frac{5}{8}=\frac{3}{8}$

b. $1-\frac{2}{3}x=\frac{9}{5}-\frac{x}{5}+\frac{3}{10}$

c. $0.7x-1.5=0.2x-8-0.4x$

116. Solve each equation.

a. $10+14x=23x-17$

b. $3x-10=5(x-4)$

c. $10-3(2x-1)=1$

d. $2x-3(x-4)=6(2x+1)$

117. Solve each equation.

a. $(x+3)(x-4)=0$

b. $3x^2-48=0$

c. $3x^2-7x=20$

d. $x^3=5x^2-6x$

118. Solve each system of equations, using the method requested.

a. Use the substitution method.

$$2x+y=8$$
$$3x-2y=4$$

b. Use the elimination method.

$$-3y=6-2x$$
$$4x+2y=10$$

119. Solve and graph:

a. $3x+5\leq 26$

b. $-9x<27$

c. $3x+7<8x-3$

d. $\frac{2}{3}-\frac{x}{5}>\frac{4}{15}$

120. Graph each equation.

a. $x=-5$

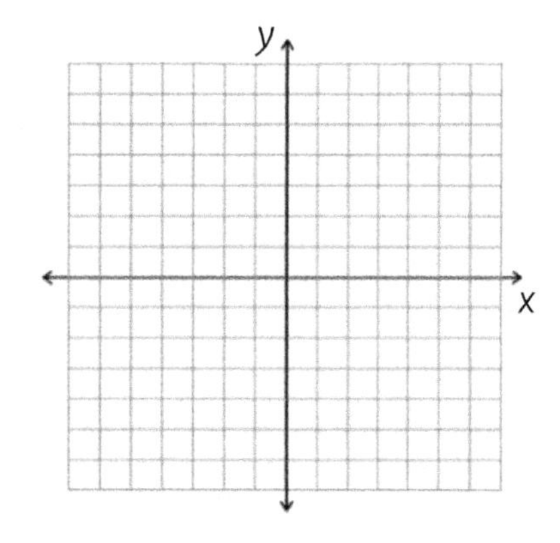

b. $y=3x-2$

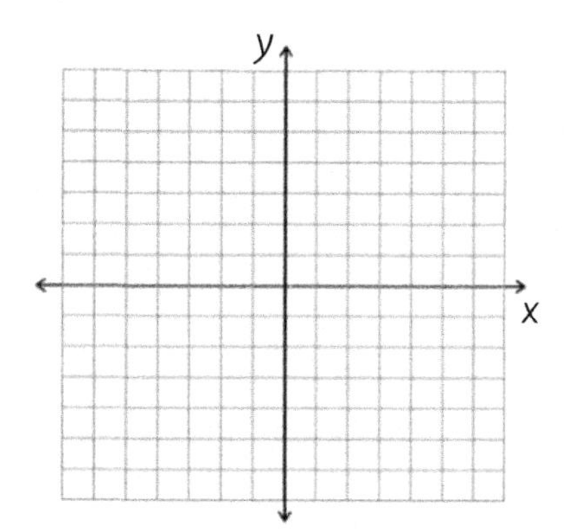

c. $4x-2y=8$

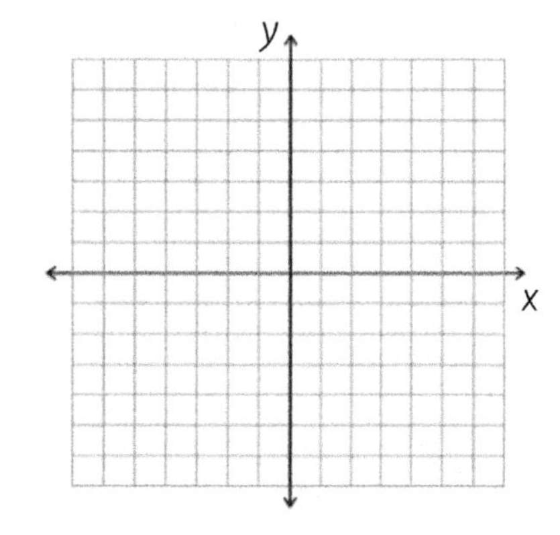

d. $y-3=-\frac{1}{3}(x+6)$

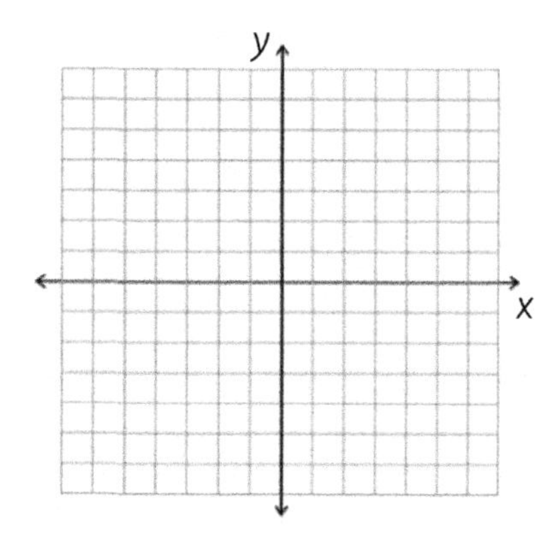

121. Graph each inequality.

a. $y > 3$

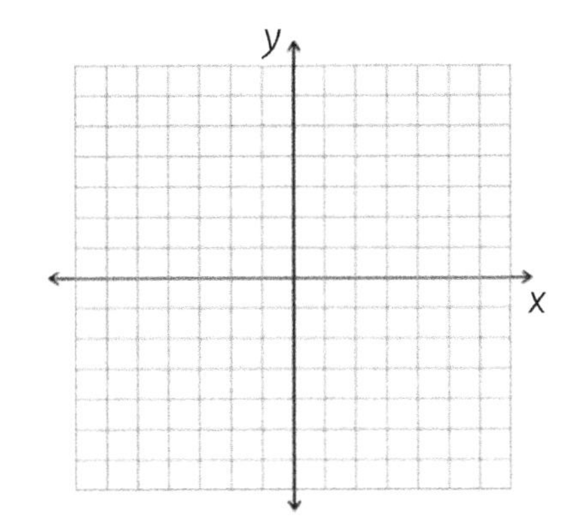

b. $y > -\frac{1}{2}x + 1$

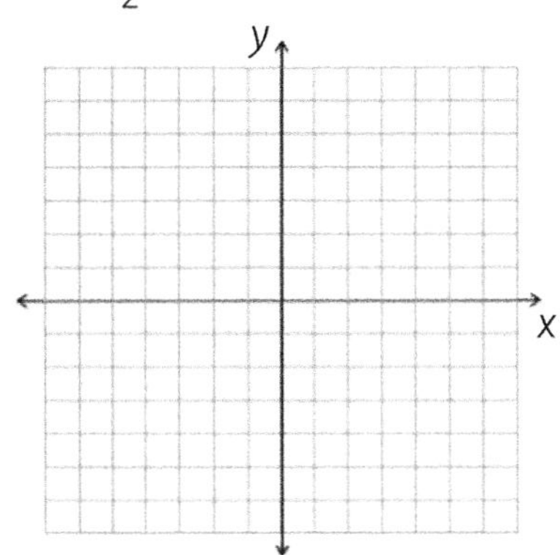

122. Graph the system of inequalities.

a. $x \geq -2$
$y < 3$

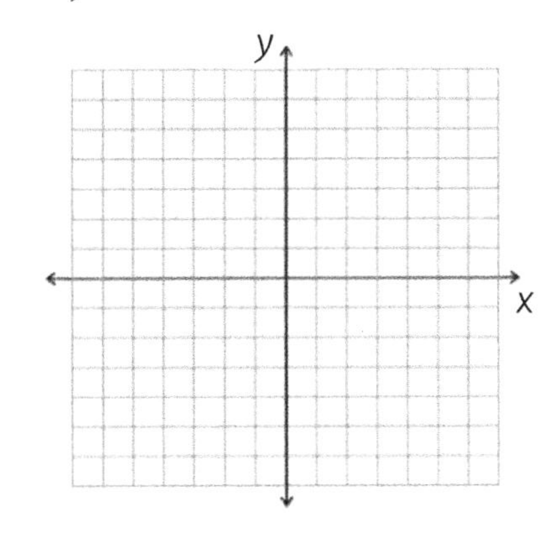

b. $y \leq 3 - x$
$2x - 5y < 5$

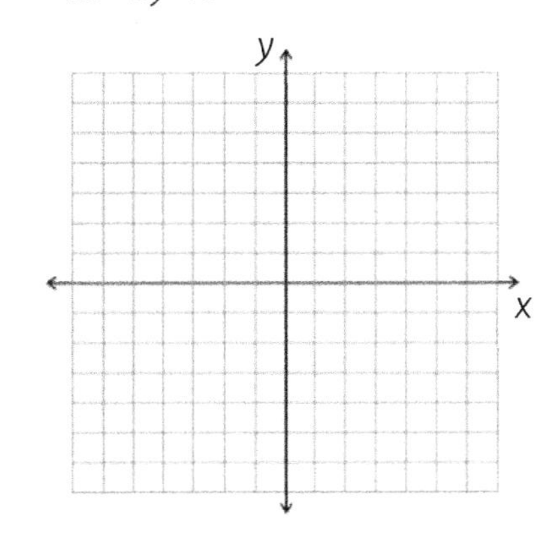

123. Graph the system of inequalities.

$x < 4$
$y > -3$
$5x - 4y > -8$

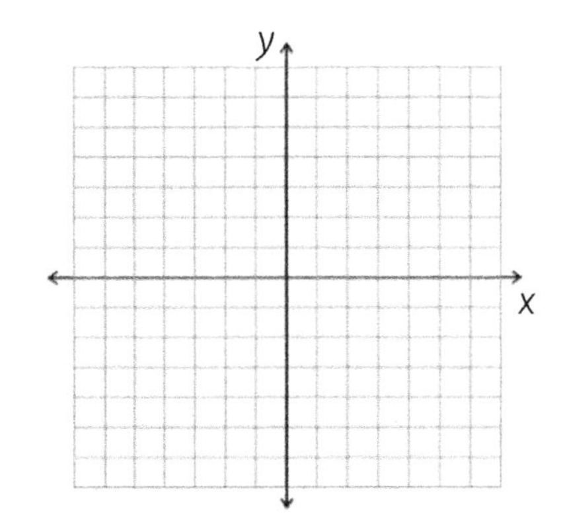

124. Simplify each expression. All exponents should be positive.

a. $y^{-3} \cdot y^{-1}$

b. $2z^{-5}$

c. $\dfrac{6x^2}{x^{-4}}$

125. Simplify each expression. All exponents should be positive.

a. $\left(3^{-1}\right)^2$

b. $\left(f^8\right)^{-2}$

c. $\left(10g\right)^{-2}$

126. Simplify each expression. All exponents should be positive.

a. $\left(-5x^{-1}\right)^2$

b. $\left(-\dfrac{y}{4}\right)^{-2}$

c. $\left(\dfrac{7}{p^5}\right)^{-4}$

127. Simplify each expression. All exponents should be positive.

a. $\dfrac{x^5}{x^{20}}$

b. $a^3 \cdot a^7 \cdot a^0$

c. $\left(x^4y^3w\right)^5$

128. Simplify each expression. All exponents should be positive.

a. $\left(-2x^3yz^2\right)^3$

b. $\left(5x^2y^3\right)^2\left(2xy\right)$

c. $\left(\dfrac{3x^3y^{-2}}{-2x^4y^5}\right)^{-2}$

129. Add, subtract, multiply or divide.

a. $(5x^3+6x^2-3x+1)+(5x^4-6x^3+2x^2-5)$

b. $(7a^2-5a+6)-(3a^2+8a-12)+(8a^2-10a+3)$

130. Multiply each expression.

a. $(2x-3y)(4x+2y)$

b. $(a^2+2b^2)^2$

c. $(x+4)(x^3-x^2-x+4)$

131. Factor each expression as much as you can.

a. x^2+7x+6

b. x^2-49

c. $10x^2-23x+12$

d. $3x^4-3$

132. Factor each expression as much as you can.

a. $8y^2-18y-5$

b. $9x^3+12x^2-45x$

c. $16x^8-81$

133. A park ranger measures a young redwood and finds that it is 35 feet tall. Several years later, he measures the tree again and it has reached a height of 175 feet. By what percent has the height increased?

134. To prepare for your math final, you spend an average of at least 30 minutes studying each night for five nights. During the first four nights, you study for 45, 35, 20, and 10 minutes, respectively. How many minutes do you need to study on the fifth night to ensure that you achieve your goal?

135. Fine Line Trucks rents an 18-foot truck for \$42 per day plus \$0.35 per mile. Judy needs a truck for one day to deliver a shipment of plants. How far can she travel and stay within her budget of \$70?

136. Your grandfather opened a stock trading account for you and put some money in the account to help you get started. You set a goal to save up enough money to buy a new laptop and started setting aside the same amount of money every week. After 5 weeks, you had \$320 in your savings account. After 11 weeks, you had \$464. How much money are you setting aside each week?

137. Find the slope of the line that passes through the points $(9,2)$ and $(-3,5)$.

138. Write the equation of the line with a slope of $\frac{1}{2}$ and a y-intercept of -2.

139. Write the equation of the line with a slope of 2 that passes through the point $(-3,5)$.

140. Write the equation of the line that contains the points $(4,-1)$ and $(-4,-3)$.

141. A school's spring festival was attended by 429 people. Admission was \$8 each for adults and \$4.50 each for children. The total receipts were \$2641. How many adults and how many children attended?

142. A restaurant charges \$5.60 for two slices of pizza and a soda. The price for three slices of pizza and three sodas is \$10.50. Determine the cost of one soda and the cost of one slice of pizza.

143. Simplify each expression as much as you can.

a. $\sqrt{48}$ b. $\sqrt{60}$ c. $\sqrt{\frac{1}{3}}$ d. $\sqrt{\frac{16}{20}}$

144. Simplify each expression as much as you can.

a. $\left(5+2\sqrt{2}\right)\left(5-2\sqrt{2}\right)$ b. $\left(\sqrt{3}+6\right)\left(\sqrt{3}-2\right)$ c. $\left(4-2\sqrt{7}\right)^2$

145. Solve each equation.

a. $\sqrt{x-1}=3$ b. $8-\sqrt{x-2}=3$ c. $\sqrt{x+7}=1+x$

146. Find the missing length.

a.

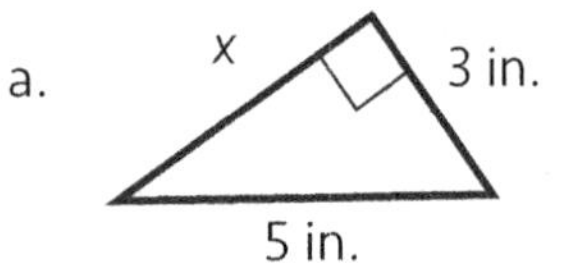

b.

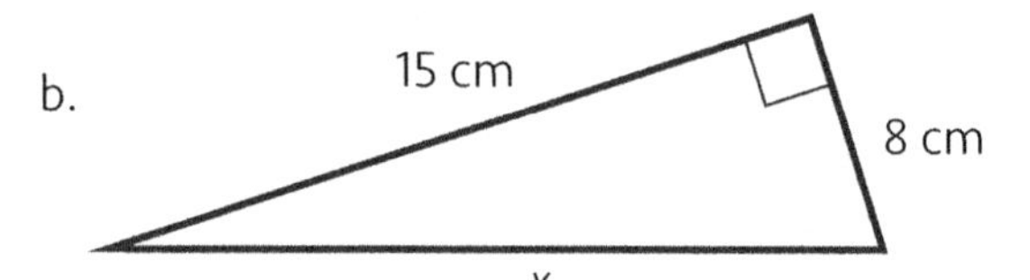

147. Determine the missing side length for each triangle shown.

a.

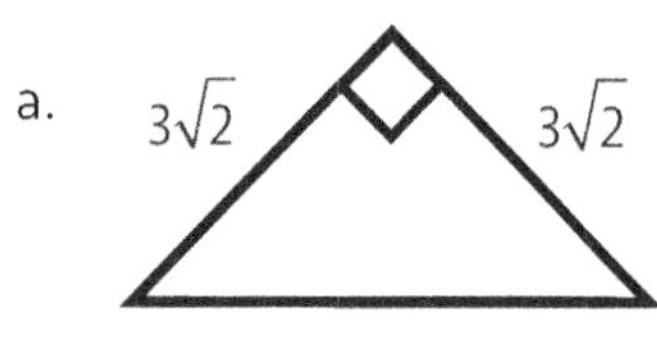

b.

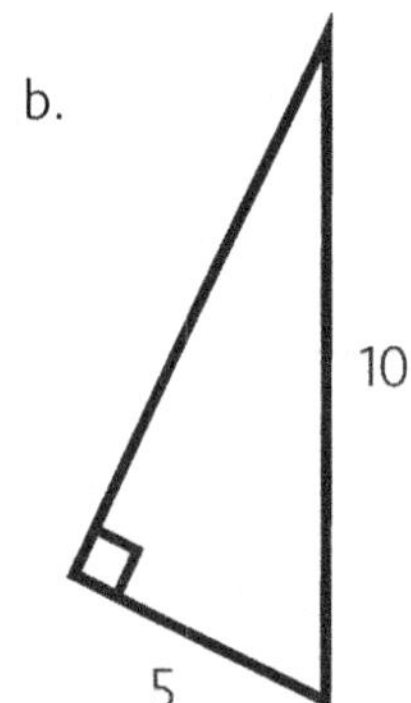

c.

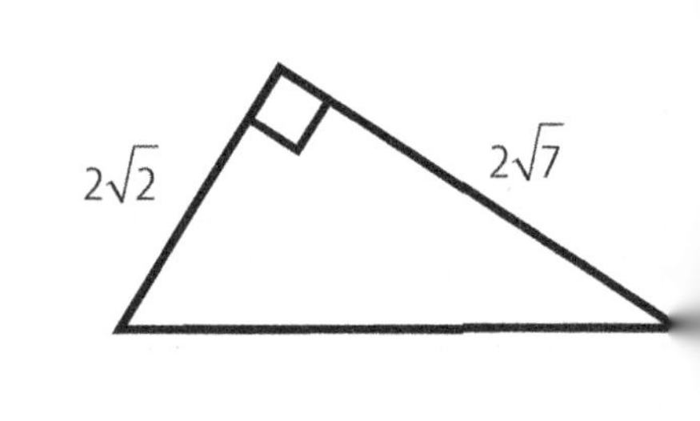

Section 8

CUMULATIVE REVIEW

148. How large is the missing angle in the figure?

a.

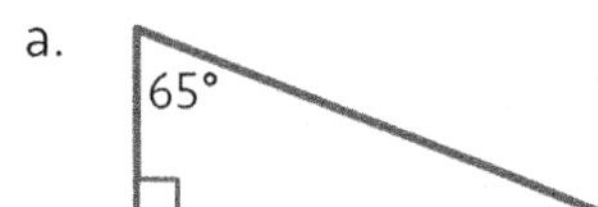

b.

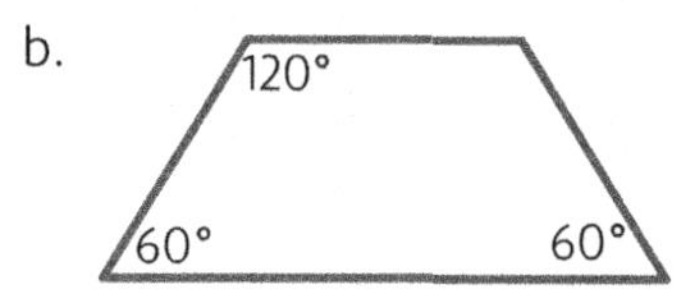

149. What is the most specific name for each figure in the previous scenario?

150. Hannah bought a car for $6,000 and she paid a sales tax of 8%. Toby bought a car for $8,000 and he paid a sales tax of 6%. Who paid a higher amount in sales taxes?

151. It takes 100 square inches of sheet metal to make a square sign. How long is the sign's perimeter?

152. Tom's baseball team scored 34 runs in their first 4 games. After the end of the 5th game, the team has scored an average of 7 runs per game. How many runs did the team score in the 5th game?

153. How many squares are in the figure below?

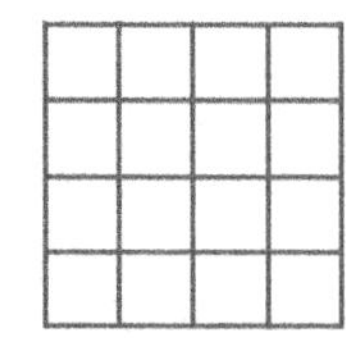

154. How many triangles are in the figure below?

155. Simplify the following expressions. Express each result using only positive exponents.

a. $2\left(3x^{-2}\right)^{-1}$

b. $\left(\dfrac{-2x^2y^3}{xy^5}\right)^{-3}$

156. Evaluate each expression in the previous scenario if $x = -2$ and $y = -1$.

157. Consider the two lines shown.

a. Identify the equation of the solid line.

b. Identify the equation of the dashed line.

c. Notice the slopes of the two lines. What is it about the slopes of these lines that makes them intersect at 90° angles?

158. Perform each operation shown below.

a. $\dfrac{1}{3}\cdot\dfrac{3}{5}$

b. $\dfrac{1}{3}\div\dfrac{3}{5}$

c. $\dfrac{1}{3}+\dfrac{3}{5}$

d. $\dfrac{1}{3}-\dfrac{3}{5}$

Section 9
ANSWER KEY

1.	a. "price" is dependent on "weight" b. "minutes" is dependent on "length" c. "pressure" is dependent on "depth"
2.	a. $P(w)$ b. $T(l)$ c. $P(d)$
3.	a. dependent: distance the car travels independent: starting height of the car b. dep.: price of the mixture indep.: weight of the mixture c. dep.: height of the plant indep.: number of days it has been growing
4.	a. $H=(0)^2+0+5 \rightarrow H=5$ b. $H=(-5)^2+(-5)+5 \rightarrow H=25$ c. t, because the value of H depends on t.
5.	a. $18 b. 3 kg c. 7 kg (the price increases by $4 per kg) d. $P = 4k$
6.	a. $2.60 b. $0.60; No, you cannot send an envelope with a weight of 0. c. 20 cents
7.	a. a little more than $3.50 (see graph below) b. 22 ounces (see graph below) Price, in dollars: 0 1 2 3 4 5 6 (22, $5) 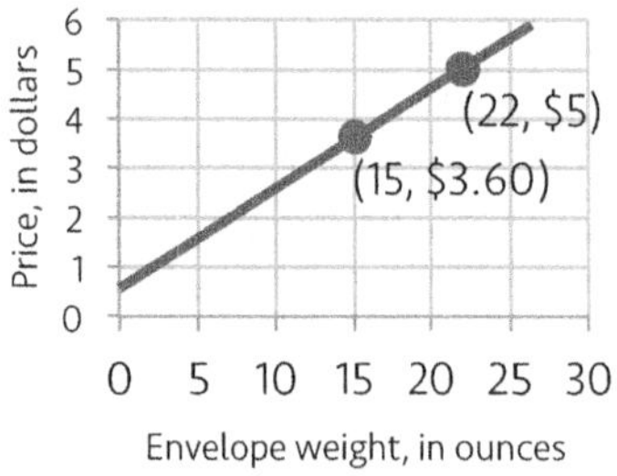 (15, $3.60) 0 5 10 15 20 25 30 Envelope weight, in ounces c. In the price function, $P = 0.6 + 0.2w$, replace P with 5 and solve for w.
8.	a. $w = 31$; $P = \$8.40$ b. $4.00
9.	a. $4.00 b. $8.60
10.	47 ounces
11.	a. $P(10)=\$2.60$ b. $P(17)=\$4.00$ c. $P(25)=\$5.60$
12.	a. $0.60+0.20(1) \rightarrow \$0.80$ b. $5=0.60+0.20w \rightarrow w=22$ ounces c. $0.60+0.20(-1) \rightarrow \$0.40$ (but this is not a meaningful number, because an envelope cannot weigh -1 ounce.
13.	a. $V(-2)=6$ b. $k = 1$
14.	a. $y=-7(-5)+11 \rightarrow y=46$ b. solve $60 = -7x + 11 \rightarrow x=-7$
15.	a. n b. T or $T(n)$ c. $T(5)=10-3(5) \rightarrow T(5)=-5$ d. solve $10 = 10 - 3n \rightarrow n=0$
16.	a. $H(-7)=-2 \rightarrow$ simplify: $\frac{-7-5}{6}$ b. $H(71)=11 \rightarrow$ solve: $\frac{y-5}{6}=11$
17.	a. 4 b. 4 c. No value. The function is not defined when $m = -7$. d. No value. The function is not defined when $m = 11$.
18.	a. $m = 0$ b. 3 values: $m = -5$, -1 and 2 c. $m = -3$, 2.75 and 3.5 d. No value. The function is not defined when $R(m)=0$.
19.	a. $f(-5)=2(-5)^2+4(-5)+3 \rightarrow f(-5)=33$ b. $9=2x^2+4x+3 \rightarrow 0=2x^2+4x-6$ $\rightarrow 0=2(x+3)(x-1) \rightarrow x=-3$ or 1
20.	a. $2,000 b. $4,000 c. 7 weeks d. $7,000
21.	a. $t = 7$ b. Andy (during the 9th week) c. Nolan (4 weeks)
22.	They both had the same rate of increase ($10,000 over 5 weeks = $2,000 per week).
23.	a. 200 dogs b. 5 and 7 (2 values)
24.	a. 2004 b. 2006
25.	a. 2003 b. About 3 years c. approx. 300 per year; the population fell from ≈1,000 to ≈400 over 2 years.
26.	decreasing 0 (zero)

27.	a. 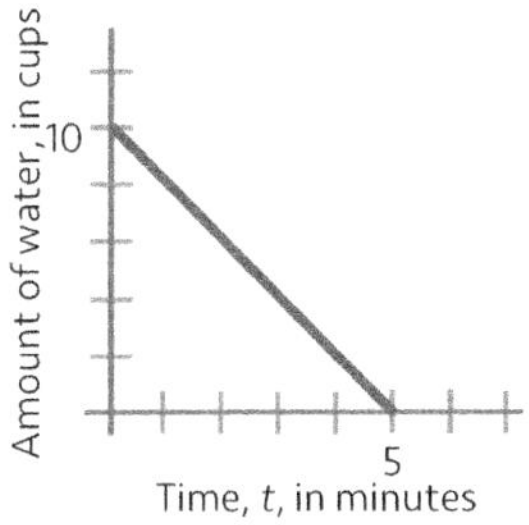 b. decreasing c. from $t = 0$ to $t = 5$ $(0 \le t \le 5)$
28.	2 cups per minute (10 cups in 5 minutes)
29.	The function is not decreasing on the interval from $x = 3$ to $x = 6$.
30.	Look for a section where the graph is horizontal.
31.	a. $-10 < x < 14$ b. $-1 \le x \le 7$
32.	$-4 < x < -1$ and $3 < x < 5$
33.	$-1 < x < 0$
34.	a. never constant b. always constant $(-\infty < x < \infty)$
35.	a. $-6 < x < -4$ and $-2 < x < 2$ and $3 < x < 7$ b. never decreasing
36.	a. no, the water level does not change during the time between one player finishing filling and the next player starting to fill b. 1 liter
37.	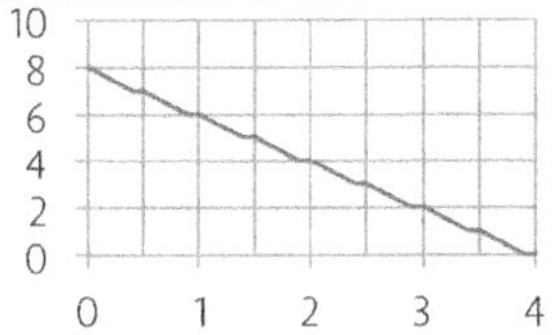 The small, repeated flat sections are drawn because the water is not running for a small amount of time after a player finishes filling a bottle, stops the water flow, and steps aside for the next player to fill.
38.	a. increasing then decreasing b. increasing then decreasing then increasing then decreasing then increasing c. All 4 segments are all increasing
39.	a – c. 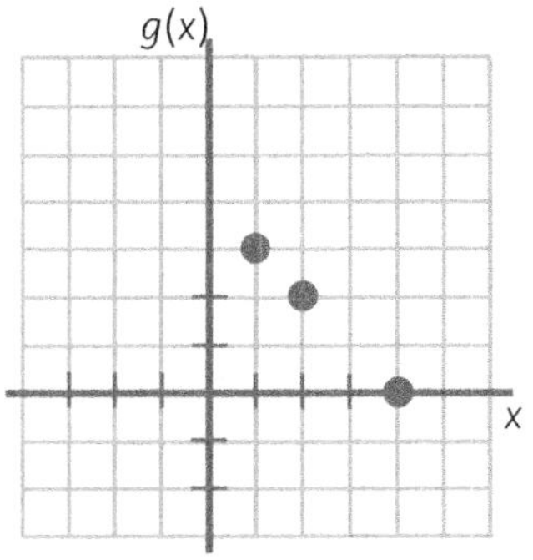 d. -6
40.	$g(x) = -x + 4$
41.	a. 2 b. 0 c. 2.5

42.	a. $x = -5$ and $x = 3$ b. $f(x)$ never equals 5 c. $x = -4, 0,$ and 5 d. $x = -3.5$ and -1.5
43.	The x-values between -3 and 3, and the x-values between 5 and 7.
44.	a. least: -5; greatest: 7 b. lowest: 0; highest: 4
45.	a. $3 + 8 = 11$ b. $g(4)$ is greater c. $-4 < x < -3$
46.	a. $-2 \le x \le 5$ b. $2 \le y \le 6$
47.	a. $-7 \le x < -2$ b. $73 \le x < \infty$
48.	a. $5 \le y < 25$ b. $-\infty < y \le 54$
49.	a. $-\infty < x < \infty$ b. $-10 \le x \le -1$
50.	a. $-\infty < y \le 2$ b. $y = -8$
51.	domain: $-\infty < x < \infty$ range: $-\infty < y < \infty$
52.	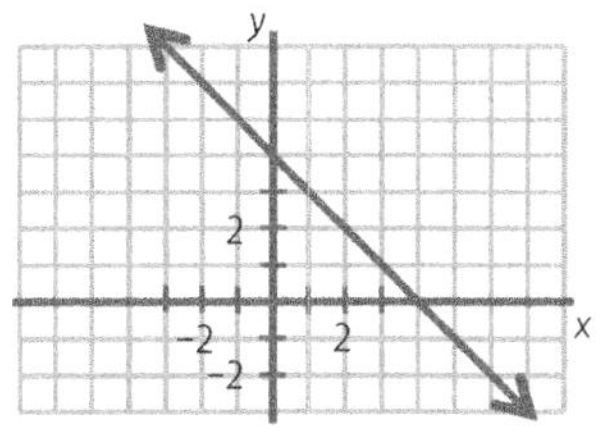 a. b. every x-value that exists (also known as "all real numbers")
53.	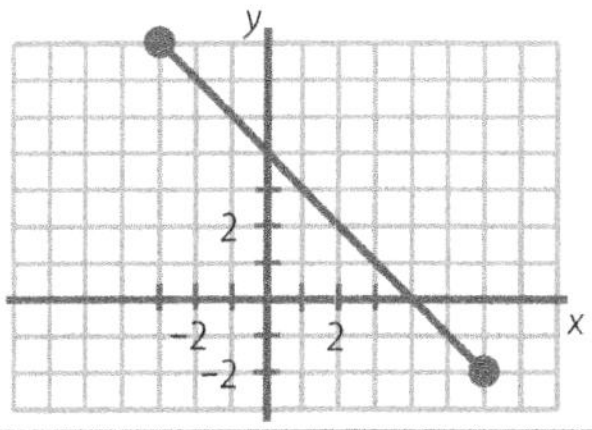
54.	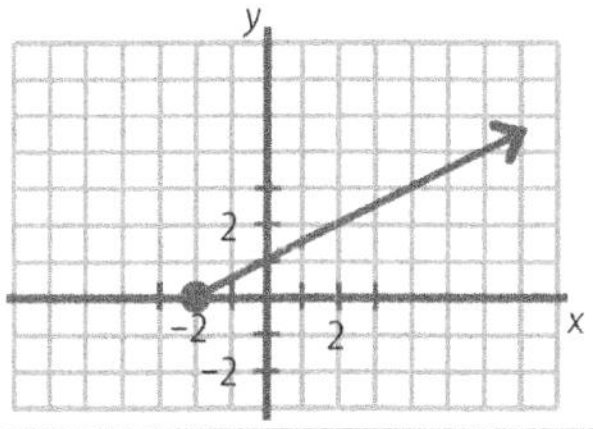
55.	Infinity is an idea, but not a specific number, so it is not a value that x can literally equal.
56.	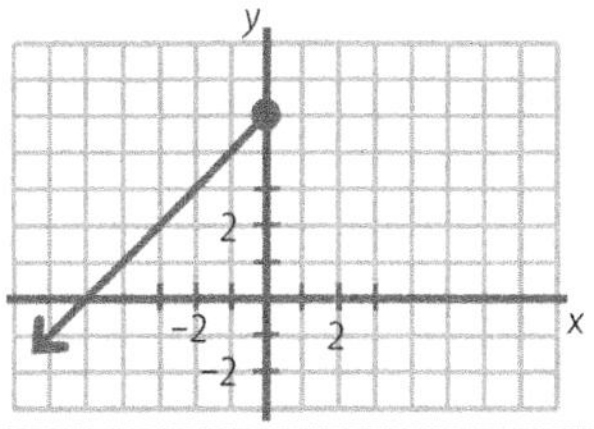
57.	$-\infty < x \le 0$
58.	ray
59.	a. d: {Freshman, Sophomore, Junior, Senior} b. r: {Orange, Yellow, Blue, Green}
60.	d: $\{-4, -1, 0, 2, 5, 7\}$ r: $\{-1, 0, 3, 6\}$
61.	range: $\{-11, -10, -9, -8, -7\}$

62.	range: $\{-3, -2, 1\}$
63.	a. -5 and 6 b. $-5 \le x \le 6$ c. 2 and 4 d. $2 \le y \le 4$
64.	$y = -x - 1$; $y = \frac{2}{3}x + 4$; $y = 4$; $y = -\frac{1}{3}x + 5$
65.	a. $-5 \le x \le -3$ b. $-3 \le x \le 0$ c. $0 \le x \le 3$ d. $3 \le x \le 6$
66.	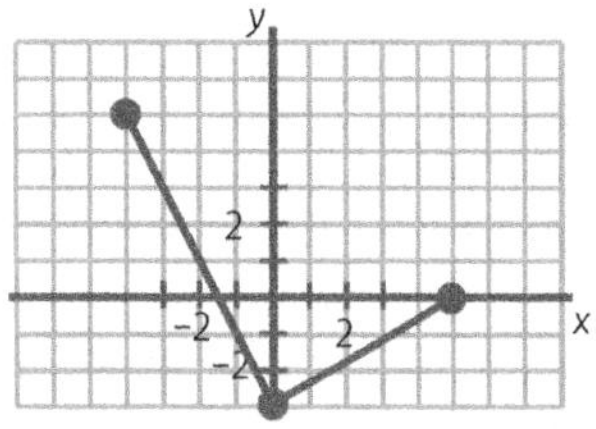 a – b. c. 0
67.	a. −4 and 5 b. $-4 \le x \le 5$ c. −3 and 5 d. $-3 \le y \le 5$
68.	a. $-6 \le x \le 7$ b. $-1 \le y \le 5$
69.	a. $0 \le t \le 90$ b. $0 \le D \le 20$
70.	3
71.	Distance, $D(t)$
72.	The vertical axis (often called the *y*-axis).
73.	a. *t* is the independent variable b. *n* is the independent variable
74.	a. $H(1.5) \approx 100$ feet b. The range should be written as $0 \le H \le 150$, because the heights of 0 feet and 150 feet are included. c. $0 \le t \le 7$ d. The height is dependent on the time
75.	(6, 2)
76.	(-5, 0)
77.	a. 30 minutes b. 550 feet
78.	a. The height rises about 150 ft in 3 min, so the height rises approx. 50 ft/min. b. The height drops approx. 50 ft/min.
79.	a. $0 \le \text{time} \le 60$ b. $0 \le \text{height} \le 550$
80.	3.75 + 2.25(8) + 0.50(5) $\rightarrow$ \$24.25
81.	a. $C(m,t) = 3.75 + 2.25m + 0.50t$ b. \$16 c. A 20-mile taxi ride, with 9 minutes of time stopped in traffic costs \$53.25.
82.	a. 10 b. 3
83.	$g(-2) = -2(-2)^2 - (-2) \rightarrow -2(4) + 2 \rightarrow -6$
84.	a. -7 b. 3 c. -17 d. 30 e. 0.5 f. 128
85.	a. 45 b. -12
86.	a. 9 b. 0 c. solve: $0 = 9 - k^2 \rightarrow k = 3, -3$ d. replace *k* with $x \rightarrow N(x) = 9 - x^2$
87.	replace *k* with "x + 1" $N(x+1) = 2(x+1)^2 + 2(x+1)$

	$N(x+1) = 2x^2 + 6x + 4$
88.	a. $0 \le t \le 7$ b. $0 \le H \le 150$ c. 75 d. the rocket is on the ground after 7 seconds
89.	a. Ellen is ahead of Anna after 2min. b. At 3min., Anna is running faster because her curve is slanting upward more than Ellen's curve. c. Ellen is behind Anna after 5min.
90.	Ellen starts faster than Anna, but she slows down and Anna' speed increases. Anna passes Ellen after 4min.
91.	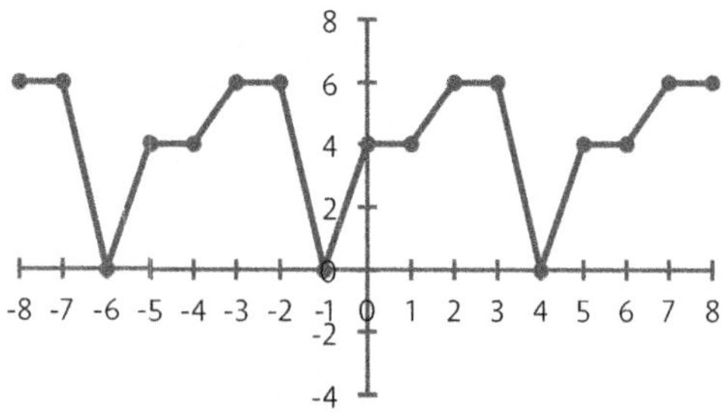
92.	a. 4; same as $f(-5), f(0), f(5)$, etc... b. 0; same as $f(4), f(9), f(14)$, etc...
93.	a. $V[R(4)] = 3$ First, find $R(4)$. $R(4) = -2(4) + 5 \rightarrow -3$ Next, find $V(-3)$. $V(-3) = 12 - (-3)^2 \rightarrow 12 - 9 \rightarrow 3$ b. $R[V(-2)] = -11$ $V(-2) = 12 - (-2)^2 \rightarrow 8$ $R(8) = -2(8) + 5 \rightarrow -11$
94.	(-4, 7)
95.	a. -4.5 b. x = -6, -2, 2 c. 5 d. x = 3
96.	a. $5(-3) \rightarrow -15$ b. First, find $g(64)$. $g(64) = 8$. Then, find $h(8)$. $h(8) = 40$. Thus, $h[g(64)] = 40$.
97.	a. $\frac{3}{5}(15) - 2 \rightarrow 9 - 2 \rightarrow 7$ b. $-4(-4) \rightarrow 16$ c. $(-3)^2 - 1 + -4\left(\frac{1}{4}\right) \rightarrow 8 + -1 \rightarrow 7$
98.	a. $f(-5) = 7 - \frac{1}{5}(-5) \rightarrow 7 + 1 \rightarrow 8$ $h(3) = (3)^2 + (3) - 2 \rightarrow 9 + 3 - 2 \rightarrow 10$ $f(-5) \cdot h(3) = 8 \cdot 10 = 80$ b. First, find $f(50) \rightarrow 7 - 10 \rightarrow -3$.

	Then, find $h(-3) \to (-3)^2 + (-3) - 2 \to 4$. Thus, $h[f(50)] = 4$. c. $6(y-2) \to 6y - 12$
99.	$G(-10) = 0$ represents the point (-10, 0). Since G is 0, the point is on the x-axis, which makes it an x-intercept.
100.	All of the points that are part of the function have x-values greater than or equal to -4 and less than or equal to 6.
101.	The domain is all of the x-values that are part of the function. The range is all of the function's y-values.
102.	Answers will vary, but the function must be inside the dashed rectangle shown below. Its highest and lowest x- and y-values must touch rectangle's edges. 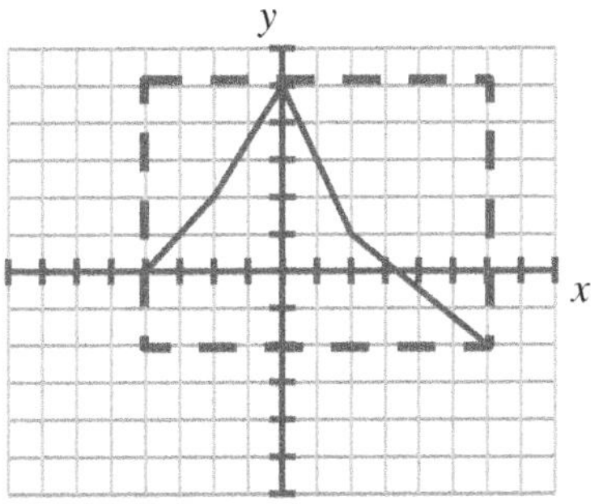
103.	a. Find all points with an x-value of -1. b. Find all points with a G-value of 2. c. range: $-1 \le G(x) \le 3$ d. domain: $-7 \le x \le 3.5$
104.	$-5 \le x \le 3 \to$ between $x=-5$ and $x=3$
105.	a. never constant b. $-\infty < x < \infty \to$ always constant
106.	$0 < t < 3$
107.	a. $H(1) \approx 50$ feet b. 2 times: t = 2 and t = 4 seconds
108.	$0 \le t \le 6$
109.	Height is the dependent quantity. The height of the ball depends on how long it has been in the air.
110.	a. 26 b. -5 c. -7
111.	a. -2 b. -43 c. 0
112.	a. 11 b. 16
113.	a. -20 b. -16 c. $-\frac{5}{3}$

114.	a. $\frac{3}{8}$ b. 6 c. -51 d. 6
115.	a. 4 b. $-\frac{33}{14}$ c. $-\frac{65}{9}$
116.	a. 3 b. 5 c. 2 d. $\frac{6}{13}$
117.	a. -3, 4 b. 4, -4 c. $-\frac{5}{3}, 4$ d. 0, 3, 2
118.	a. $\left(\frac{20}{7}, \frac{16}{7}\right)$ b. $\left(\frac{21}{8}, -\frac{1}{4}\right)$
119.	a. $x \le 7$ b. $x > -3$ c. $x > 2$ d. $x < 2$
120.	a. 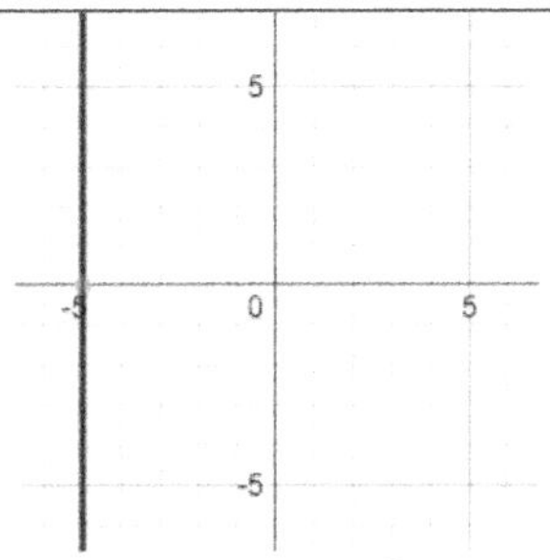 b. 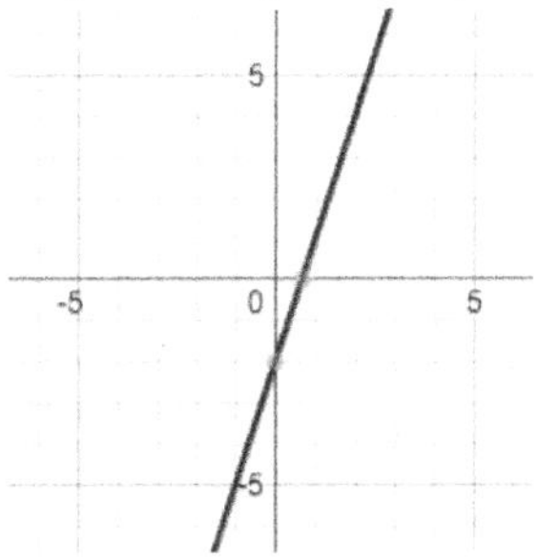 c. 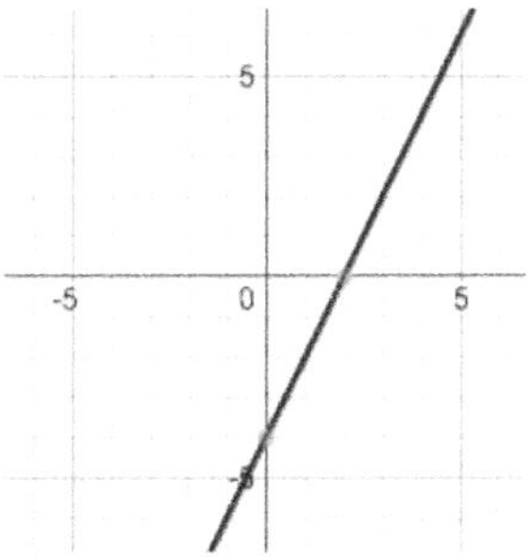 d. 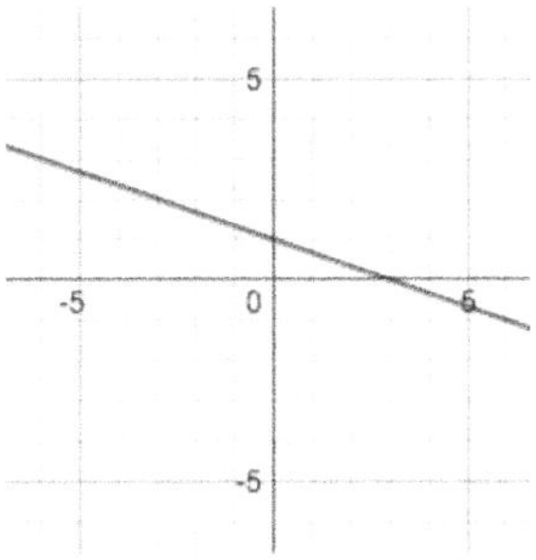

121.	a. 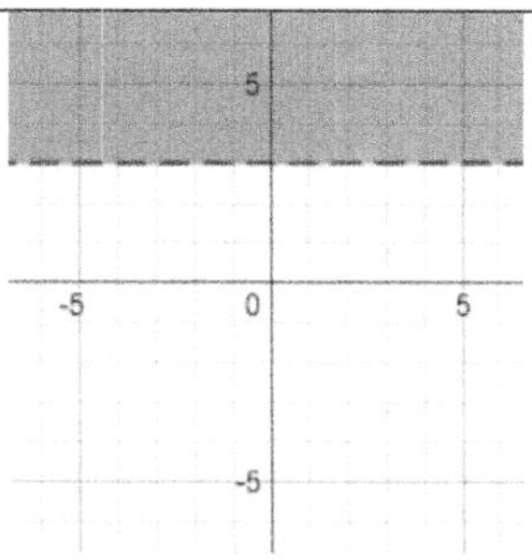 b. 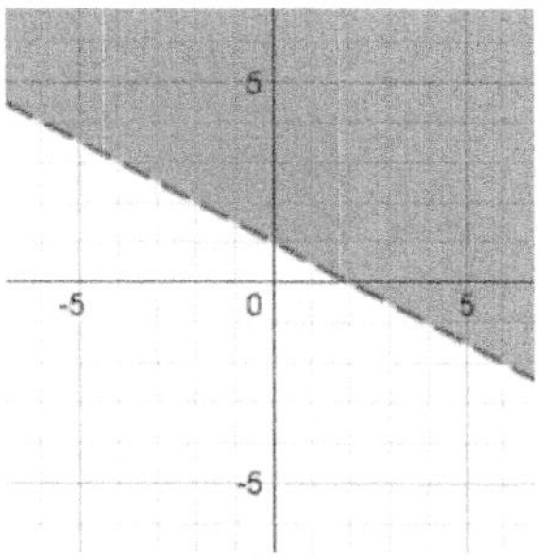
122.	a. 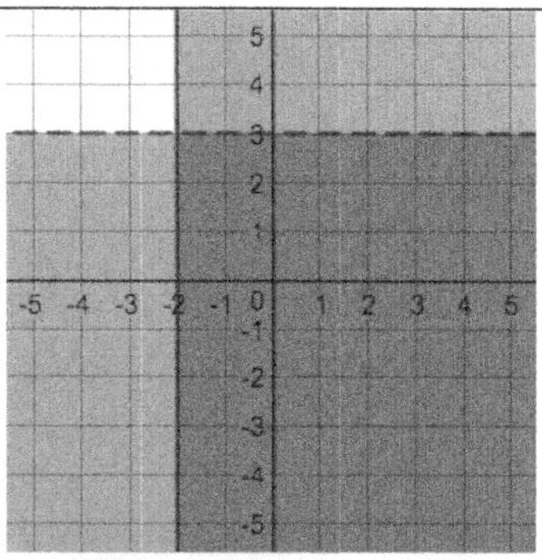 b. 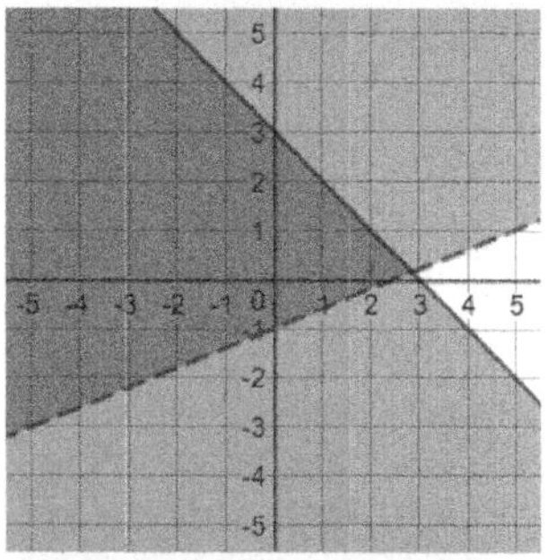
123.	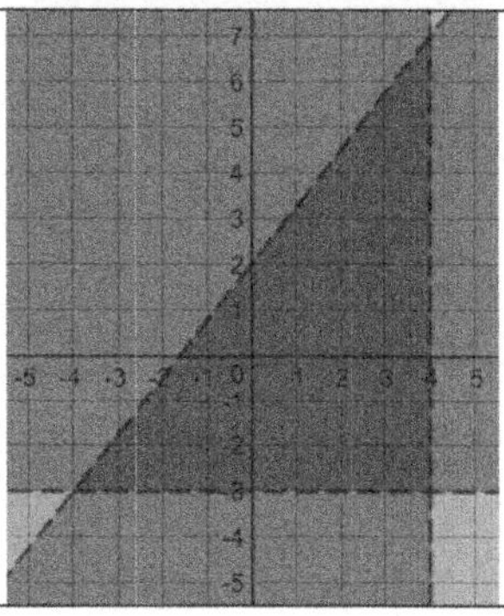
124.	a. $\frac{1}{y^4}$ b. $\frac{2}{z^5}$ c. $6x^6$
125.	a. $\frac{1}{3^2} \rightarrow \frac{1}{9}$ b. $\frac{1}{f^{16}}$ c. $\frac{1}{100g^2}$
126.	a. $\frac{25}{x^2}$ b. $\frac{16}{y^2}$ c. $\frac{p^{20}}{7^4}$
127.	a. $\frac{1}{x^{15}}$ b. a^{10} c. $x^{20}y^{15}w^5$
128.	a. $-8x^9y^3z^6$ b. $50x^5y^7$ c. $\frac{4x^2y^{14}}{9}$
129.	a. $5x^4-x^3+8x^2-3x-4$ b. $12a^2-23a+21$
130.	a. $8x^2-8xy-6y^2$ b. $a^4+4a^2b^2+4b^4$ c. $x^4+3x^3-5x^2+16$
131.	a. $(x+6)(x+1)$ b. $(x+7)(x-7)$ c. $(5x-4)(2x-3)$ d. $3(x^2+1)(x+1)(x-1)$
132.	a. $(4y+1)(2y-5)$ b. $3x(3x-5)(x+3)$ c. $(4x^4+9)(2x^2+3)(2x^2-3)$
133.	400% increase
134.	$x \geq 40$ minutes
135.	$m \leq 80$ miles
136.	$24.00 per week
137.	$m=-\frac{1}{4}$
138.	$y=\frac{1}{2}x-2$
139.	$y=2x+11$
140.	$y=\frac{1}{4}x-2$
141.	203 adults 226 children
142.	pizza = $2.10 soda = $1.40
143.	a. $4\sqrt{3}$ b. $2\sqrt{15}$ c. $\frac{\sqrt{3}}{3}$ d. $\frac{2\sqrt{5}}{5}$
144.	a. 17 b. $4\sqrt{3}-9$ c. $44-16\sqrt{7}$
145.	a. 10 b. 27 c. 2
146.	a. 4 in. b. 17 cm
147.	a. 6 b. $5\sqrt{3}$ c. 6
148.	a. 25° (The sum of the 3 angles of a triangle is 180°) b. 120° (The sum of the 4 angles of a quadrilateral is 360°)
149.	a. scalene right triangle b. isosceles trapezoid
150.	They paid the same amounts.

	$.08(6{,}000) = .06(8{,}000) = \$480$
151.	40 inches (The sides of the square are 10 inches long.)
152.	They scored 1 run. (Since the team's average after 5 games is 7 runs per game, the team has scored 35 runs overall after 5 games.)
153.	30 squares. (16 1x1 squares; 9 2x2 squares; 4 3x3 squares; 1 4x4 square)
154.	10 triangles (5 of each type shown below)
155.	a. $2\left(\frac{3}{x^2}\right)^{-1} \rightarrow 2\left(\frac{x^2}{3}\right) \rightarrow \frac{2x^2}{3}$ b. $\left(\frac{-2x}{y^2}\right)^{-3} \rightarrow \left(\frac{y^2}{-2x}\right)^{3} \rightarrow \frac{y^6}{(-2)^3 x^3} \rightarrow \frac{y^6}{-8x^3}$
156.	a. $\frac{2(-2)^2}{3} \rightarrow \frac{2 \cdot 4}{3} \rightarrow \frac{8}{3}$ or $2\frac{2}{3}$ b. $\frac{(-1)^6}{-8(-2)^3} \rightarrow \frac{1}{-8 \cdot -8} \rightarrow \frac{1}{64}$
157.	a. $y = -2x - 2$ b. $y = \frac{1}{2}x + 4$ c. The lines have opposite reciprocal slopes of $-\frac{2}{1}$ and $\frac{1}{2}$, so they are perpendicular.
158.	a. $\frac{3}{15} \rightarrow \frac{1}{5}$ b. $\frac{1}{3} \div \frac{3}{5} \rightarrow \frac{1}{3} \cdot \frac{5}{3} \rightarrow \frac{5}{9}$ c. $\frac{1}{3} \cdot \frac{5}{5} + \frac{3}{5} \cdot \frac{3}{3} \rightarrow \frac{5}{15} + \frac{9}{15} \rightarrow \frac{14}{15}$ d. $\frac{1}{3} \cdot \frac{5}{5} - \frac{3}{5} \cdot \frac{3}{3} \rightarrow \frac{5}{15} - \frac{9}{15} \rightarrow -\frac{4}{15}$

Made in the USA
Las Vegas, NV
03 December 2022